The Art of Micro Frontends

Build highly scalable, distributed web applications with multiple teams

Florian Rappl

‹packt›

The Art of Micro Frontends

Group Product Manager: Kaustubh Manglurkar

Publishing Product Manager: Bhavya Rao

Book Project Manager: Aishwarya Mohan

Senior Editor: Debolina Acharyya

Technical Editor: Simran Ali

Copy Editor: Safis Editing

Indexer: Rekha Nair

Production Designer: Jyoti Kadam

DevRel Marketing Coordinators: Anamika Singh and Nivedita Pandey

First published: June 2021
Second edition: September 2024

Production reference: 1010824

Published by Packt Publishing Ltd.
Grosvenor House
11 St Paul's Square
Birmingham
B3 1RB, UK

ISBN 978-1-83546-035-1

www.packtpub.com

To my family for their continuous support and dedication. My life would not be so colorful without you, and it would be hard to keep following the right path. Thanks for your guidance and dedication!

– Florian Rappl

Foreword

I have known and worked with *Florian Rappl* for the last eight years. Florian is one of these rare architects who also loves to write code. Over these eight years, I've witnessed many projects where Florian took the lead architecture role – introducing a micro frontend solution and guiding teams to a successful implementation. He is one of the world's .NET and micro frontend community leading experts. In this book, *The Art of Micro Frontends*, Florian takes you on a learning journey that covers implementation ideas, scalability aspects, organizational issues, as well as security aspects. This will help you to successfully launch your own micro frontend solution – either from scratch or using one of the established frameworks.

For a long time, even large-scale systems have been developed as so-called monoliths, where the entire functionality has been implemented as a single and tightly integrated solution. For the backend part, the architecture pattern of microservices has become a very popular choice to address some of the downsides of a monolithic solution, such as strong coupling of components, complex deployment scenarios, or large code bases managed by a single development team. Nowadays, microservice architectures build the foundation for many modern backend implementations.

Only recently, similar architectural approaches have found their way into frontend solutions. One of the key requirements for flexible and extendable user interfaces is the modularization of components, especially for scenarios in which layout and available features strongly depend on the user context, or contained frontends for microservices need to be exposed seamlessly as a unified application.

Decoupling functionality blocks is already a broadly applied aspect of modern monolithic frontend solutions, but together with the freedom to select the appropriate technology and the ability to independently develop in smaller teams, the micro frontend architecture pattern replicates further key benefits of microservice architecture in the frontend. In particular, the capability of loading individual modules dynamically into the frontend enables the development, testing, and provisioning of features in shorter and self-contained release cycles. This makes the architecture pattern of micro frontends an excellent choice for large-scale and complex web developments.

As for selecting microservices as the basis for backend architecture, applying the approach of micro frontends might not be the ideal and canonical choice for every frontend solution. The isolation and decoupling of functional building blocks of the user interface introduces another layer of complexity; hence, a well-designed and modularized monolithic web application might be a valuable or even a better fit for a given usage scenario. It is essential to evaluate and assess the advantages and disadvantages upfront as part of the solution concept. In this context, this book will support you in the process of designing the right frontend solution architecture.

Become a developer superhero and build scalable web applications that leverage micro frontends.

Lothar Schöttner

Founder and CEO of smapiot, previously Solution Architect for Accenture and Microsoft.

Contributors

About the author

Florian Rappl is a solution architect working on distributed web applications for digital transformation and IoT projects. His main interest lies in the implementation of micro frontends and their impact on teams and business models. In this area, he led several teams, realizing many successful projects over the last few years. As the lead architect, he has helped to create outstanding web applications for many industry-leading companies. He regularly gives lectures on software design patterns and web development. Florian has won multiple prizes for his work over the years and is recognized as a Microsoft MVP for development technologies. He started his career in software engineering before studying physics and helping to build an energy-efficient supercomputer. Florian currently lives in Munich, Germany, with his wife and their two daughters.

About the reviewers

Norton Almeida brings over 15 years of experience as a senior frontend developer to his role as a technical reviewer. Previously a consultant for Fortune 500 companies, Norton transitioned to a project in 2020 that utilized Vue.js and a rudimentary implementation of a micro frontend architecture. This experience sparked a deep interest in the subject, leading him to become an enthusiastic researcher and reviewer of technical literature.

Sunil Raj Thota is a seasoned software engineer with extensive experience in web development and AI applications. Currently working at Amazon QuickSight Team, Sunil has previously contributed to significant projects at Yahoo Inc., enhancing user engagement and satisfaction through innovative features at Yahoo and AOL Mail. He has also worked at Northeastern University as a research assistant and at MOURI Tech as a senior software engineer, optimizing multiple websites and leading successful project deployments. Sunil co-founded ISF Technologies, where he championed user-centric design and agile methodologies. He has also contributed to the book *Demystifying AI for Web Development*. His academic background includes a master's in analytics from Northeastern University and a bachelor's in electronics and communications engineering from Andhra University.

Table of Contents

Preface   xvii

Part 1: The Hive – Introducing Frontend Modularization

1

Why Micro Frontends?   3

Technical requirements	3	Emerging web standards	10
The evolution of web applications	4	Isolation via web components	11
Programming the web	4	Frame communication	11
The social web	5	Web workers and proxies	13
The separation of the frontend and the backend	6	**Faster TTM**	**14**
Everything becomes micro	**7**	Decreasing onboarding time	14
From SOA to microservices	7	Multiple teams	15
The advantages of microservices	8	Isolated features	16
The disadvantages of microservices	9	A/B testing	16
Micro and frontend	10	**Summary**	**17**

2

Common Challenges and Pitfalls   19

Technical requirements	19	Security	23
Performance	20	Central user management	23
Resource caching	20	Individual user management	24
Bundle size	21	Script execution	25
Request optimizations	22	**Knowledge sharing**	**26**

Reliability   28   Pattern libraries   29

User experience   29   **Summary**   **30**

Wording   29

3

Deployment Scenarios   31

Technical requirements   **32**   Using dedicated pipelines   39

Central deployments   **32**   **Hybrid solutions**   **40**

Using a monorepo   32   Scheduled releases   40

Joining multiple repositories   35   Triggering on change   40

Distributed deployments   **37**   **Summary**   **41**

Using a monorepo   37

4

Domain Decomposition   43

Technical requirements   **44**   Functional split   49

Principles of DDD   **44**   Example decomposition   51

Modules   44   **Architectural boundaries**   **54**

Bounded context   45   Shared capabilities   55

Context map   46   Choosing the right level of freedom   56

Strategic domain design versus tactical design 46   Accessing the DOM   57

                                                           Universality of micro frontends   58

Separation of concerns   **48**   **Summary**   **59**

Technical split   48

Part 2: Dry Honey – Implementing Micro Frontend Architectures

5

Types of Micro Frontend Architectures   63

Technical requirements   **64**   Static versus dynamic micro frontends   65

The micro frontend landscape   **64**   Horizontal versus vertical split   69

Backend- versus frontend-driven
micro frontends   70

Summary   74

6

The Web Approach   75

Technical requirements   76
Basics of the web approach   76
The architecture   77
Sample implementation   77
Potential enhancements   80

Advantages and disadvantages
of the web approach   80
Using links for navigation   81

Central linking directory   81
Local linking directory   82
Using fragments with iframes   83
Security   83
Accessibility   83
Layout   83
Summary   85

7

Server-Side Composition   87

Technical requirements   88
Basics of server-side composition   88
The architecture   89
Sample implementation   89
Potential enhancements   100

Advantages and disadvantages of
server-side composition   100
Reviewing Mosaic 9   101
Introducing Podium   101
Known users   102

Creating a composition layout   102
Understanding layout responsibilities   103
Using SSI   103
Using ESI   104
Using JS template strings   104

Setting up micro frontend projects   106
Podlets   107
Examining the lifecycle   108

Using a dedicated rendering server   108
Summary   110

8

Edge-Side Composition   111

Technical requirements   112
Basics of edge-side composition   112

The architecture   112
Sample implementation   113

Potential enhancements   117

**Advantages and disadvantages of
edge-side composition**   118

SSI and ESI   121

SSI   121
ESI   122

Stitching in BFFs   124
Summary   125

9

Client-Side Composition   127

Technical requirements   128
Basics of client-side composition   128

The architecture   128
Sample implementation   130
Potential enhancements   135

**Advantages and disadvantages of
client-side composition**   135
Diving into web components   136

Understanding web components   136
Isolating styles with shadow DOM   137

**Composing micro frontends
dynamically**   138

Using a micro frontend discovery service   138
Updating micro frontends at runtime   139

Summary   141

10

SPA Composition   143

Technical requirements   144
Basics of SPA composition   144

The architecture   144
Sample implementation   145
Potential enhancements   150

**Advantages and disadvantages of SPA
composition**   150
Building a core SPA shell   151

Activating pages   151
Sharing dependencies   153

Integrating SPA micro frontends   156

Declaring the lifecycle   156
Using cross-framework components   157

Exploring communication patterns   159

Exchanging events   160
Sharing data   161
Extending components   162

**Optimizing hydration and
progressive rendering**   163
Summary   165

11

Siteless UIs   167

Technical requirements	168	Creating a siteless UI runtime	181
The basics of siteless UIs	168	Building a runtime with Piral	182
The architecture	168	Deploying a runtime with Piral	184
Sample implementation	170	**Writing siteless UI modules**	**185**
Potential enhancements	178	Looking at a pilet's life cycle	185
Advantages and disadvantages of siteless UIs	**179**	Implementing framework-agnostic components	188
Comparing siteless UIs and serverless	**179**	**Implementing islands composition with Qwik**	**190**
Developing locally	180	**Summary**	**194**
Publishing modules	181		

Part 3: Bee Brood – Implementation Details

12

Sharing Dependencies with Module Federation   197

Technical requirements	197	**Understanding Native Federation**	**205**
Sharing dependencies between micro frontends	198	**Achieving independence with SystemJS**	**208**
Utilizing Module Federation	200	Summary	210

13

Isolating CSS   211

Technical requirements	211	Using the shadow DOM	217
Understanding the consequences of open styling	212	Using modern CSS features for isolation	219
Implementation techniques to scope CSS	213	Summary	222

14

Securing the Application   231

Technical requirements	223	Limiting script access	228
Using web standards to harden		Verifying authenticity	229
security	224	Summary	230

15

Decoupling Using a Discovery Service   231

Technical requirements	231	Using advanced capabilities	237
Avoiding hidden monoliths	232	Summary	240
Implementing a discovery service	234		

Part 4: Busy Bees – Scaling Organizations

16

Preparing Teams and Stakeholders   245

Technical requirements	246	Handling product owners and	
Communicating with C-level		steering committees	249
stakeholders	246	Team organization	250
Managing expectations	246	Understanding possible team setups	251
Writing executive summaries	248	Changing team organizations	254
		Summary	255

17

Dependency Management, Governance, and Security   257

Technical requirements	258	Sandboxing micro frontends	263
Sharing all or nothing	258	General security concerns and	
What about change management?	259	mitigations	265
Establishing a governance model	261	Summary	268

18

Impact of Micro Frontends on UX and Screen Design   269

Technical requirements	270	Sharing designs efficiently	276
Always adding one	270	Creating designs without designers	278
Learning to start at zero	274	Summary	279

19

Building a Great Developer Experience   281

Technical requirements	282	Improving testability	288
Providing a minimum developer experience	282	Achieving the best developer experience	288
Supporting development in standard IDEs	282	Integrating error codes	289
Improving the scaffolding experience	283	Providing an offline-first development environment	290
Establishing a decent developer experience	285	Customizing via browser extensions	292
Centralizing code documentation	285	Implementing a developer portal	292
Using videos	286	Summary	294
Assisting with code analysis	287		

20

Case Studies   295

Technical requirements	295	Impact	303
A user-facing portal solution	295	A healthcare management solution	304
Problem description	296	Problem description	304
Team setup	296	Team setup	304
Solution	297	Solution	305
Impact	298	Impact	307
An administration portal solution	300	An e-commerce solution	309
Problem description	300	Problem description	309
Team setup	300	Team setup	309
Solution	300	Solution	309

Impact   311   Solution   312

An application for mobile banking   311   Impact   314

Problem description   312   **Summary   315**

Team setup   312   **Epilogue   315**

Index   317

Other Books You May Enjoy   328

Preface

The pattern of micro frontends is a web architecture for frontend development, borrowed from the idea of microservices in software development, where each module of a frontend is developed and shipped in isolation to avoid complexity and a single point of failure for your frontend.

By definition, micro frontends are the technical representation of a business subdomain. They allow independent implementations with the same or different technology and they should minimize the code shared with other subdomains and are owned by autonomous teams. In this book micro frontends are explored in full detail – starting from simple systems up to complex organizations.

To give you the broadest idea what micro frontends are about the book is divided into four parts – each covering a different angle. The first part only covers the theoretical background, while the second part is all about the most commonly used architecture patterns. The third part contains solutions to the practical challenges. In the final part organizational topics are discussed to help you set up a micro frontend solution in a variety of scenarios.

Who this book is for

This book is for solution architects, software architects, developers, and frontend engineers. Basic frontend knowledge of HTML and CSS with a decent knowledge of JavaScript and its ecosystem, including Node.js with npm, is assumed.

What this book covers

Chapter 1, Why Micro Frontends?, covers micro frontends in general, their primary areas of use, as well as the challenges and problems they bring. We will explore what strategies can be used to mitigate these.

Chapter 2, Common Challenges and Pitfalls, discusses the most important challenges and pitfalls when implementing micro frontends, together with a path leading to a proper solution.

Chapter 3, Deployment Scenarios, looks at the scalability of micro frontends with respect to their deployment. This chapter includes CI/CD pipeline examples and their ideal use cases.

Chapter 4, Domain Decomposition, explores a way of thinking that can be used when deciding what should be placed in a micro frontend. This chapter introduces methods from domain-driven design that can be used to make these decisions.

Chapter 5, Types of Micro Frontend Architectures, introduces the phase space to create micro-frontend architecture, including the most popular patterns available. This chapter outlines the advantages and disadvantages of the extremes within the phase space.

Chapter 6, The Web Approach, discusses the simplest approach to handling micro frontends by leveraging existing web technologies, such as iframes or links.

Chapter 7, Server-Side Composition, discusses a popular backend method of combining frontend fragments that come from different servers into a single website, as well as its extension into an islands-based architecture.

Chapter 8, Edge-Side Composition, looks at an even more simplified method than server-side composition to compose a website on the edge, using a reverse proxy setup.

Chapter 9, Client-Side Composition, shows how to leverage web components to compose one website from different fragments in a user's browser.

Chapter 10, SPA Composition, discusses a way of bringing together different SPA websites in a joint solution, composed within a user's browser.

Chapter 11, Siteless UIs, introduces a micro frontend pattern that brings the popular properties of serverless functions to the frontend.

Chapter 12, Sharing Dependencies with Module Federation, deep-dives into sharing dependencies between different micro frontends, using technologies such as Module Federation or Native Federation.

Chapter 13, Isolating CSS, introduces several techniques to prevent styling collisions when placing fragments from different micro frontends in a single website.

Chapter 14, Securing the Application, contains useful strategies to mitigate potential vulnerabilities, using established web standards and implementation guidelines.

Chapter 15, Decoupling Using a Discovery Service, discusses the need and usefulness of a central service to aggregate different micro frontends.

Chapter 16, Preparing Teams and Stakeholders, deals with the organizational shift that is necessary when introducing micro frontends.

Chapter 17, Dependency Management, Governance, and Security, provides some guidance on dependency sharing and general micro frontend governance for projects of any kind. This chapter also touches on the topic of security, from deployment to runtime.

Chapter 18, Impact of Micro Frontends on UX and Screen Design, reveals the most critical aspects that need to be handled when creating designs for micro frontend solutions with practically unlimited scalability.

Chapter 19, Building a Great Developer Experience, lists the most crucial properties to include to satisfy internal or external developers of a project. This is crucial to keep up a high level of productivity.

Chapter 20, Case Studies, lists five different real-world micro frontend projects with their background, core decisions, and the overall taken architecture.

To get the most out of this book

All the examples in the book have been created with simplicity in mind. They all work similarly and only require knowledge of core frontend technologies, such as JavaScript with HTML and CSS. The tooling to make the code run also uses the most popular frontend toolchain based on Node.js. As such, if you know how to work with JavaScript and have used Node.js with npm before, you'll have no problems following the examples presented in the book.

Software/hardware covered in the book	Operating system requirements
Node.js 20 (or higher)	Windows, macOS, or Linux (any distro)
npm 10 (or higher)	Git
ECMAScript 2020 (or higher)	

For the code in Chapter 8, you'll need to run Docker. Alternatively, you can just set up a local nginx server, although that will be a lot more complicated than just running the provided Docker file.

If you are using the digital version of this book, we advise you to type the code yourself or access the code from the book's GitHub repository (a link is available in the next section). Doing so will help you avoid any potential errors related to the copying and pasting of code.

Download the example code files

You can download the example code files for this book from GitHub at `https://github.com/PacktPublishing/The-Art-of-Micro-Frontends-Second-Edition`. If there's an update to the code, it will be updated in the GitHub repository.

We also have other code bundles from our rich catalog of books and videos available at `https://github.com/PacktPublishing/`. Check them out!

Code in Action

The Code in Action videos for this book can be viewed at `https://packt.link/yNAAE`

Conventions used

There are a number of text conventions used throughout this book.

`Code in text`: Indicates code words in text, database table names, folder names, filenames, file extensions, pathnames, dummy URLs, user input, and Twitter handles. Here is an example: "One of the browser's security guarantees is that an `HttpOnly` cookie will never be read out by a script."

A block of code is set as follows:

```
const generalProxy = new Proxy(() => generalProxy, {
  get(target, name) {
    if (name === Symbol.toPrimitive) {
      return () => ({}).toString();
    } else {
      return generalProxy();
    }
  },
});
```

When we wish to draw your attention to a particular part of a code block, the relevant lines or items are set in bold:

```
# Initialize a new Node.js project
npm init -y
# Add the http-server package to the dependencies
npm i http-server --save
```

Any command-line input or output is written as follows:

```
npx pilet debug
```

Bold: Indicates a new term, an important word, or words that you see on screen. For instance, words in menus or dialog boxes appear in **bold**. Here is an example: "As an example, when **micro frontend 2 (MF2)** from *Figure 15.2* is loaded, it has to register **Component B** in the **Component Registry**."

> **Tips or important notes**
> Appear like this.

Get in touch

Feedback from our readers is always welcome.

General feedback: If you have questions about any aspect of this book, email us at customercare@packtpub.com and mention the book title in the subject of your message.

Errata: Although we have taken every care to ensure the accuracy of our content, mistakes do happen. If you have found a mistake in this book, we would be grateful if you would report this to us. Please visit www.packtpub.com/support/errata and fill in the form.

Piracy: If you come across any illegal copies of our works in any form on the internet, we would be grateful if you would provide us with the location address or website name. Please contact us at copyright@packtpub.com with a link to the material.

If you are interested in becoming an author: If there is a topic that you have expertise in and you are interested in either writing or contributing to a book, please visit authors.packtpub.com.

Share Your Thoughts

Once you've read *The Art of Micro Frontends*, we'd love to hear your thoughts! Scan the QR code below to go straight to the Amazon review page for this book and share your feedback.

https://packt.link/r/1835460356

Your review is important to us and the tech community and will help us make sure we're delivering excellent quality content.

Download a free PDF copy of this book

Thanks for purchasing this book!

Do you like to read on the go but are unable to carry your print books everywhere?

Is your eBook purchase not compatible with the device of your choice?

Don't worry, now with every Packt book you get a DRM-free PDF version of that book at no cost.

Read anywhere, any place, on any device. Search, copy, and paste code from your favorite technical books directly into your application.

The perks don't stop there, you can get exclusive access to discounts, newsletters, and great free content in your inbox daily

Follow these simple steps to get the benefits:

1. Scan the QR code or visit the link below

https://packt.link/free-ebook/978-1-83546-035-1

2. Submit your proof of purchase
3. That's it! We'll send your free PDF and other benefits to your email directly

Part 1:
The Hive – Introducing
Frontend Modularization

In this part, you will not only gain a deeper understanding of micro frontends and their primary applications, but also get to grips with their various challenges and the strategies to mitigate them.

This part covers the following chapters:

- *Chapter 1, Why Micro Frontends?*
- *Chapter 2, Common Challenges and Pitfalls*
- *Chapter 3, Deployment Scenarios*
- *Chapter 4, Domain Decomposition*

1

Why Micro Frontends?

Every journey needs to start somewhere. Quite often, we'll find ourselves looking for a solution to a particular problem. Knowing the problem in depth is usually part of the solution. Without a detailed understanding of the various challenges, finding a proper way to handle the situation seems impossible.

After almost three decades of web development, we've reached a point where anything is possible. The boundaries between classical desktop applications, mobile native applications, and websites have been torn apart. These days, many websites are actually web applications and offer a tool-like character.

As developers, we find ourselves in a quite difficult situation. Very often, the cognitive load to handle the underlying complexities of code bases is way too high. Consequently, inefficiencies and bugs stalk us.

One possible way out of this dilemma is the architectural pattern of micro frontends. In this chapter, we'll discover the reasons why micro frontends have become quite a popular choice for projects of any kind.

We will cover the following key topics in this chapter:

- The evolution of web applications
- Everything becoming micro
- Emerging web standards
- Faster **time to market** (TTM)

Without further ado, let's get right into the first topic.

Technical requirements

To follow the code samples in this book, you need knowledge of JavaScript and how to use the command line. You should have Node.js (version 20 or higher) installed, using the instructions at `https://nodejs.org`.

For this chapter, there is no source code and no **Code in Action** (**CiA**) video available.

The evolution of web applications

Before looking for reasons to use micro frontends, we should look at why micro frontends came to exist at all. How did the web evolve from a small **proof of concept (POC)** running on a NeXT computer in a small office at **Conseil Européen pour la Recherche Nucléaire (CERN)/The European Organization for Nuclear Research** to become a central piece of the information age?

Programming the web

My first contact with web development was in the mid-1990s. Back then, the web was mostly composed of static web pages. While some people were experienced enough to bring in some dynamic websites using the **Common Gateway Interface (CGI)** technology, as shown in *Figure 1.1*, most webmasters did not have knowledge of this or wanted to spend money on **server-side rendering (SSR)**. The term *webmaster* was commonly used for somebody who was in charge of a website and instead of doing SSR, everything was crafted by hand upfront.

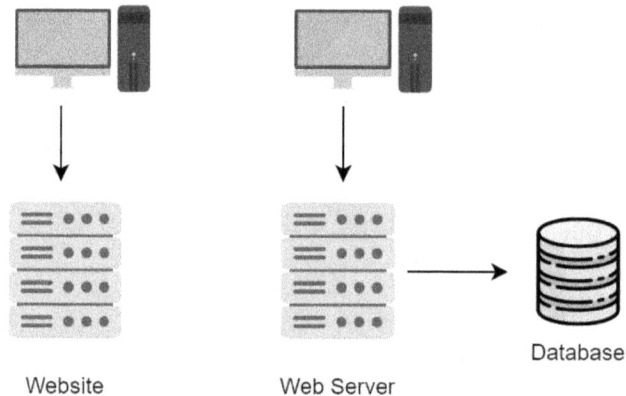

Figure 1.1 – The web changes from static to dynamic pages

To avoid duplication and potential inconsistencies, a new technology was used – frames declared in `<frameset>` tags. Frames allowed websites to be displayed within websites. Effectively, this enabled the reuse of things such as a menu, header, or footer on different pages. While frames have been removed from the **HyperText Markup Language 5 (HTML5)** specification, they are still available in all browsers. Their successor still lives on today – the **inline frame (iframe)** tag, `<iframe>`.

One of the downsides of frames was that link handling became increasingly difficult. To get the best performance, the right target had to be selected explicitly. Another difficulty was encountered in correct **Uniform Resource Locator (URL)** handling. Since navigation was only performed on a given frame, the displayed page address did not change.

Consequently, people started to look for alternatives. One way was to use SSR without all the complexity. By introducing some special HTML comments as placeholders containing some server instructions, a generic layout could be added that would be dynamically resolved. The name for this technique was **Server Side Includes**, more commonly known as **SSI**.

The Apache web server was among the first to introduce SSI support through the `mod_ssi` module. Other popular web servers, such as Microsoft's **Internet Information Services (IIS)**, followed quickly afterward. The progressive nature and Turing completeness of the allowed instructions made SSI an instant success.

It was around that time that websites became more and more dynamic. To tame the CGI, many solutions were implemented. However, only when a new programming language called **PHP: Hypertext Preprocessor (PHP)** was introduced did SSR become mainstream. The cost of running PHP was so cheap that SSI was almost forgotten.

The social web

With the rise of Web 2.0 and the capabilities of JavaScript – especially for dynamic data loading at runtime, known as **asynchronous JavaScript and XML (AJAX)** – the web community faced another challenge. SSR was no longer a silver bullet. Instead, the dynamic part of websites had to live on the client – at least partially. JavaScript became much more important, and the complexity of dividing an application into multiple areas (build, server, and client), along with their testing and development, skyrocketed.

As a result, frameworks for **client-side rendering (CSR)** emerged. The first-generation frameworks such as Backbone.js, Knockout.js, and AngularJS all came with architecture choices similar to the popular frameworks for SSR. They did not intend to place the full application on the client, and they did not intend to grow indefinitely.

In practice, this looked quite different. The application size increased and a lot of code was served to the client that was never intended to be used by the current user. Images and other media were served without any optimizations, and the web became slower and slower.

Of course, we've seen the rise of tools to mitigate this. While JavaScript minification has become nearly as old as JavaScript itself, other tools for image optimization and **Cascading Style Sheets (CSS)** minification came about to improve the situation too.

The missing link was to combine these tools into a single pipeline. Thanks to Node.js, the web community received a truly magnificent gem. Now, we had a runtime that could not only bring JavaScript to the server side but also allow the use of cross-platform tooling. New task runners such as Grunt and Gulp started making frontend code easier to develop efficiently.

The Web 2.0 movement also made the reuse of web services directly from the UI running in the browser popular, as shown in the following figure.

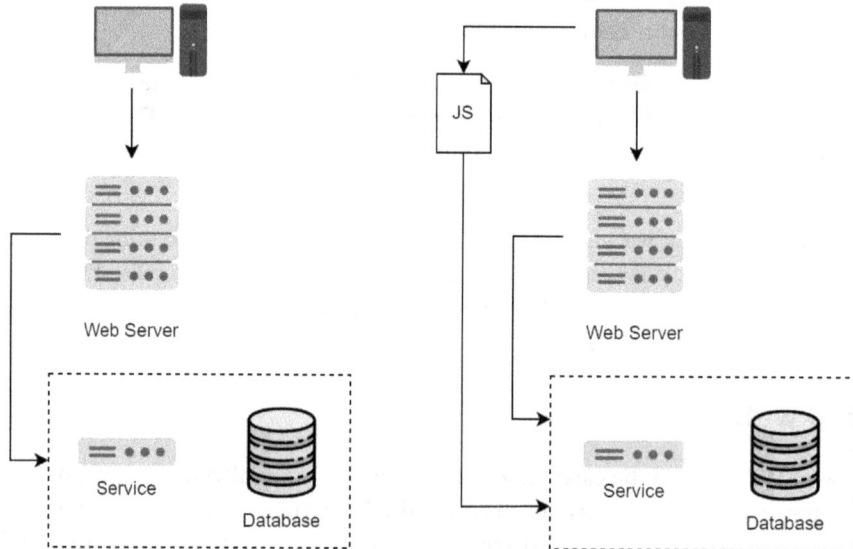

Figure 1.2 – With the Web 2.0 movement, services, and AJAX enter common architectures

Having dedicated backend services makes sense for multiple use cases. First, we can leverage AJAX in the frontend to do partial reloads, as outlined in *Figure 1.2*. Another use case is to allow other systems to access the information too. This way, useful data can be monetarized as well. Finally, the separation between representation (usually using HTML) and structure (using formats such as **Extensible Markup Language (XML)** or **JavaScript Object Notation (JSON)**) allows reuse across multiple applications.

The separation of the frontend and the backend

The enhanced rendering capabilities on the frontend also accelerated the separation between the backend and the frontend. Suddenly, there was no giant monolith that handled user activities, page generation, and database queries in one code base. Instead, the data handling was put into an **Application Programming Interface (API)** layer that could be used for page generation in SSR scenarios or directly from code running in the user's browser. In particular, applications that handle all rendering in the client were labeled a **single-page application (SPA)**.

Such a separation not only brought benefits – the design of these APIs became an art on their own. Providing suitable security settings and establishing a great performance baseline became more difficult, however. Ultimately, this also became a challenge from a deployment perspective.

From a user experience perspective, the capability of doing partial page updates is not without challenges either. Here, we rely on indicators such as loading spinners, skeleton styles, or other methods to transport the right signals to the user. Another thing to take care of here is correct error handling. Should we retry? Inform the user? Use a fallback? Multiple possibilities exist; however, besides making a good decision here, we also need to spend some time implementing and testing it.

Nevertheless, for many applications, the split into a dedicated frontend and a dedicated backend part is definitely a suitable choice. One reason is that dedicated teams can work on both sides of the story, thus making the development of larger web applications more efficient.

The gain in development efficiency, as we will see, is a driving force behind the move toward micro frontends. Let's see how the trend toward modularization got started in web development.

Everything becomes micro

In the previous section, we discovered that the eventual split into dedicated parts for the frontend and backend of an application may be quite useful. Staying on the backend, the rise of smaller services that need to be orchestrated and joined together can be observed across the industry in the form of so-called microservices.

From SOA to microservices

While the idea of a **service-oriented architecture (SOA)** is not new, what microservices brought to the table is freedom of choice. In the early 2000s, the term SOA was introduced to already try that out. However, for SOA, we saw a broad palette of requirements and constraints. From the communicating protocol and the discoverability of the API to the consuming applications, everything was either predetermined – or at least strongly recommended.

With the ultimate failure of SOA, the community tried to stay away from building too many services too quickly. In the end, it was not only about the constraints imposed by SOA but also about the orchestration struggle of deploying multiple services with potential interdependencies. How could we go ahead and reliably deploy multiple services into – most likely – two or three environments?

When Docker was introduced, it solved a couple of problems. Most importantly, it brought about a way to ensure deployments worked reliably. Combined with tools such as Ansible, releasing multiple services seemed feasible and a possible improvement. Consequently, multiple teams around the world started breaking existing boundaries. The common design choices used by these teams formed the basis of a general architecture style, known as microservices.

When using microservices, there was only a single recommendation – build services that are responsible for one thing. The method of communication (usually **representational state transfer (REST)** with JSON) was not determined. Whether services should have their own databases or not was left to the developer. Whether services should communicate directly or indirectly with each other could be decided by the architect. Ultimately, best practices for many situations emerged, cementing the success of microservices even more.

While some purists see microservices as *fine-grained SOA*, most people believe in the freedom of choice of using only the **single-responsibility principle (SRP)**. The following figure shows web applications using microservices with a SPA.

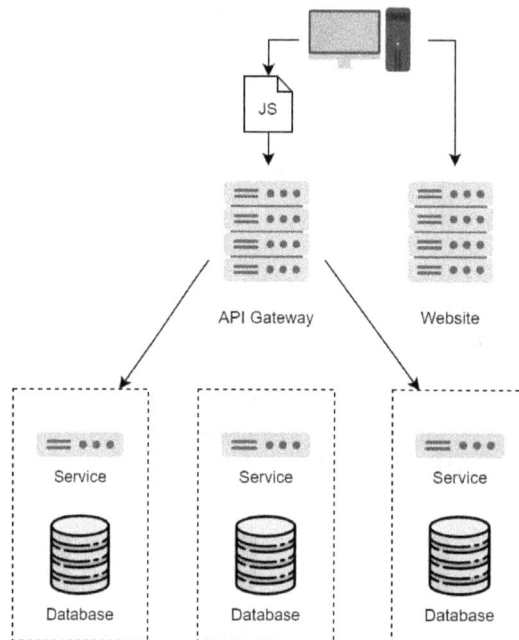

Figure 1.3 – Current state-of-the-art web applications using microservices with a SPA

In keeping with people building backends using microservices, frontends needed to follow a similar dynamic approach. Consequently, a lot of applications are heavily built on top of JavaScript, leveraging SPAs to avoid having to deal with the complexities of having to define both SSR and CSR. This is also shown in *Figure 1.3*.

The advantages of microservices

Why should a book on micro frontends spend any time looking at the history of microservices or their potential advantages? Well, for one, both start with *micro*, which could be a coincidence but actually is not. Another reason is that, as we will see, the background and history of both architectures are actually quite analogous. Since microservices are older than micro frontends, they can be used to give us a view into the future, which can only be helpful when making design decisions.

How did microservices become the de facto standard for creating backends these days? Sure, a good monolith is still the right choice for many projects but is not commonly regarded as a popular choice. Most larger projects immediately head for the microservices route, independent of any evaluations or other constraints they could encounter.

Besides their obvious popularity, microservices also come with some intrinsic advantages, outlined as follows:

- Failures only directly concern a single service
- Multiple teams can work independently
- Deployments are smaller
- Frameworks and programming languages can be chosen freely
- The initial release time is smaller
- They have clear architectural boundaries

Every project willingly choosing microservices hopes to benefit from one or more of the advantages from this list, but there are – of course – some disadvantages too.

The disadvantages of microservices

For every advantage, there is also a disadvantage. What really matters is our perspective and how we weigh the arguments. For instance, if failures only concern a single service, this results in increased complexity for debugging. A service that depends on another service that just failed will also behave strangely. Finding the root cause for why another service failed and working out how to mitigate this situation is increasingly difficult.

Nevertheless, if we go for microservices, we should value their advantages over their disadvantages. If we cannot live with the disadvantages, then we need to either look for a different pattern or try our best to mitigate the disadvantage in question as much as possible.

Another example of where an advantage might lead to a disadvantage is in the freedom of choice in terms of frameworks and programming languages. Depending on the number of developers, a certain number of languages may be supported as well. Anything that exceeds that number will be unmaintainable. It's certainly quite fashionable to try new languages and frameworks; however, if no one else is capable of maintaining a new service written in some niche language, we'll have a hard time scaling the development.

Some of the most well-known disadvantages are outlined as follows:

- Increased orchestration complexity
- Multiple points of failure
- Debugging and testing becoming more difficult
- A lack of responsibility
- Eventual inconsistencies between the different services
- Versioning hell

This list may look intimidating at first, but we need to remember that developers have already addressed these issues with best practices and enhanced tooling. Ultimately, it's our call to make the best of the situation and choose wisely.

Micro and frontend

Microservices sound great, so applying their principles to the frontend seems to be an obvious next step. The easiest solution would be to just implement microservices that serve HTML instead of JSON. Problem solved, right? Not so fast…

Just assembling the HTML would not be so difficult, but we need to think bigger than that. We need assets such as JavaScript, CSS, and image files. How can we interleave JavaScript content? If global variables are placed with the same name, which one wins? Should the URLs of all the files resolve to the individual services serving them?

There are more questions than answers here, but one thing is sure – while doing so would be possible in theory, in practice the experience would be limited, and we would lose the separation between the backend and the frontend, which – as we found out – is fundamental to today's mindset of web development.

Consequently, micro frontends did not exist until quite recently, when some of these barriers could be broken. As we'll see, there are multiple ways to establish a micro frontend solution. Similar to microservices, there are no fixed constraints. Ultimately, a project uses micro frontends when independent teams can freely deploy smaller chunks of the frontend without having to update the main application.

Besides some technical reasons, there is also another argument why micro frontends were established later than microservices – they were not required beforehand.

As outlined, the web has only recently seen an explosion in frontend code. Beforehand, either separated pages with no uniform user experience were sufficient or a monolith did the job sufficiently well. With applications growing massively in size, micro frontends came to the rescue.

Server-side solutions such as SSI or dedicated logic can help a fair bit to establish a micro frontend solution, but what about bringing this dynamic composition to the client? As it turns out, emerging web standards make this possible too.

Emerging web standards

There are a couple of problems that make developing micro frontends that run on the client problematic. Among other considerations, we need to be able to isolate code as much as possible to avoid conflicts. However, in contrast to backend microservices, we may also want to share resources such as a runtime framework; otherwise, exhausting the available resources on a user's computer is possible.

While this sounds exotic at first, performance is one of the bigger challenges for micro frontends. On the backend, we can give every service a suitable amount of hardware. On the frontend, we need to live with the browser running code on a machine that the user chooses. This could be a larger desktop machine, but it could also be a Raspberry Pi or a smartphone.

One area where recent web standards help quite a bit is with style isolation. Without style isolation, the **Document Object Model (DOM)** would treat every style as global, resulting in accidental style leaks or overwrites due to conflicts.

Isolation via web components

Styles can be isolated using the technique of a shadow DOM, which is part of the web components specification. A shadow DOM enables us to write components that are projected into the parent DOM without being part of the parent DOM. Instead, only a container element (referred to as the host) is directly mounted. Not only do style rules from the parent not leak into the shadow DOM; but style definitions in general are not applied either.

Consequently, a shadow DOM requires style sheets from the parent to be loaded again – if we want to use these styles. This isolation thus only makes sense if we want to be completely autonomous regarding styling in the shadow DOM.

The same isolation does not work for the JavaScript parts. Through global variables, we are still able to access the exports of the parent's scripts. Here, another drawback of the web components' standard shines through. Usually, the way to transport shadow DOM definitions is by using custom elements. However, custom elements require a unique name that cannot be redefined.

A way around many of the JavaScript drawbacks is to fall back to the mentioned `<iframe>` element. An iframe comes with style and script isolation, but how can we communicate between the parent and the content living in the frame? Let's find out.

Frame communication

Originally introduced in HTML5, the `window.postMessage` function has proven to be quite useful when it comes to micro frontends. This was already introduced quite early in reusable frontend pieces. Today, we can find reusable frontend pieces that rely on frame communication – for instance, in most chatbot services or cookie consent services.

While sending a message is one part of the story, the other part is to receive messages. For this, a `message` event was introduced on the `window` object. The difference between this event and many other events is that this one also gives us an `origin` property, which allows us to track the URL of the sender.

The URL of a particular frame to send a message to is also necessary when sending a message. Let's see an example of such a process.

Here is the HTML code of the parent document:

```
<!doctype html>
<iframe src="iframe.html" id="iframe"></iframe>
<script>
setTimeout(() => {
iframe.contentWindow.postMessage('Hello!', '*');
}, 1000);
</script>
```

After a second, we'll send a message containing the `Hello!` string to the document loaded from `iframe.html`. For the URL, we use the `*` wildcard string.

The iframe can be defined like so:

```
<!doctype html>
<script>
window.addEventListener('message', event => {
  const text = document.body.appendChild(
    document.createElement('div'));
  text.textContent = ` Received "${event.data}" from
    ${event.origin}`;
});
</script>
```

This will display the posted message in the frame.

> **Important note**
>
> The access and manipulation of frames are only forbidden when using a cross-origin. The definition of a cross-domain origin is one of the security fundamentals that the web is currently built upon. An origin consists of the protocol (for example, **HTTPS**), the domain, and the port. Different subdomains also correspond to different origins. Many HTTP requests can only be performed if the cross-origin rules are positively evaluated by the browser. Technically, this is referred to as **Cross-Origin Resource Sharing (CORS)**. See the following *MDN Web Docs* page for more on this: `https://developer.mozilla.org/en-US/docs/Web/HTTP/CORS`.

While these communication patterns are quite fruitful, they do not allow you to share any resources. Remember that only strings can be transported via the `postMessage` function. While these strings may be complex objects serialized as JSON, they can never be true objects with nested cyclic references or functions inside.

An alternative is to give up on direct isolation and instead work on indirect isolation. A wonderful possibility is to leverage web workers to do this.

Web workers and proxies

A web worker represents an easy way to break out of the single-threaded model of the JavaScript engine, without all the usual hassle of multithreading.

As with iframes, the only method of communication between the main thread and the worker thread is to post messages. A key difference, however, is that the worker runs in another global context that is different from the current `window` object. While some APIs still exist, many parts are either different or not available at all.

One example is in the way of loading additional scripts. While standard code may just append another `<script>` element, the web worker needs to use the `importScripts` function. This one is synchronous and allows not only one URL to be specified but also multiple ones. The given URLs are loaded and evaluated in order.

So far so good, but how can we use the web worker if it comes with a different global context? Any frontend-related code that tries to do DOM manipulation will fail here. This is where proxies come to the rescue.

A proxy can be used to capture desired object accesses and function calls. This allows us to forward certain behaviors from the web worker to the host. The only drawback is that the `postMessage` interface is asynchronous in nature, which can be a challenge when synchronous APIs should be mimicked.

One of the simplest proxies is actually the *I can handle everything* proxy. This small amount of code, as shown in the following snippet, provides a solid basis for almost all stubs:

```
const generalProxy = new Proxy(() => generalProxy, {
  get(target, name) {
    if (name === Symbol.toPrimitive) {
      return () => ({}).toString();
    } else {
      return generalProxy();
    }
  },
});
```

The trick here is that this allows it to be used like `generalProxy.foo.bar().qxz`, without any problems and with no undefined access or invalid functions.

Using a proxy, we can mock necessary DOM APIs, which are then forwarded to the parent document. Of course, we would only define and forward API calls that could be safely used. Ultimately, the trick is to filter against a safe list of acceptable APIs.

When facing the problem of transporting object references such as callbacks from the web worker to the parent, we may fall back to wrapping these. A custom marker may be used to allow the web worker to be called to run the callback at a later point in time.

We will go into the details of such a solution later – for now, it's enough to know that the web offers a fair share of security and performance measures that can be utilized to allow strong modularization.

For the final section, let's close out the original question this chapter asks and answers, *why micro frontends?* with a look at the business reasons for choosing them.

Faster TTM

As previously mentioned, there are technical and business reasons why micro frontends have been a bit more difficult to fully embrace than their backend counterparts. My personal opinion is that the technical reasons are significant, but they are not as crucial as the business reasons.

When implementing micro frontends, we should never look for technical reasons alone. Ultimately, our applications are created for end users. As a result, end users should not be impacted negatively by our technical choices.

But not only should the user experience be a driver from a business perspective – the productivity level and the development process should also be taken into consideration. Ultimately, this impacts the user experience just as much and represents the basis for any implementation – how fast we can ship new features or adjust to some user surveys.

Decreasing onboarding time

Twenty years ago, most developers that entered a company stayed for quite a while. Over time, this decreased to around two years for most companies (https://hackerlife.co/blog/san-francisco-large-corporation-employee-tenure). The average employment duration is shown in *Figure 1.4.*

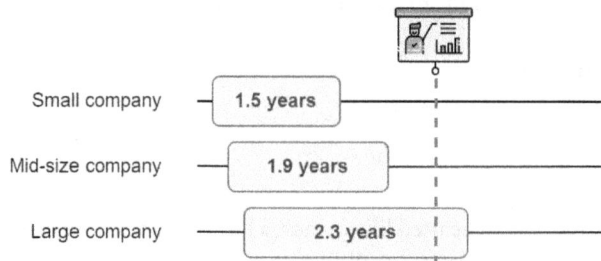

Figure 1.4 – The average time of developers at a single company

If a standard developer gets productive in a larger code base after six months, this means that a large fraction of the investment (about 25%) is not really fulfilled. Being productive from day one should be the goal, even if it is unrealistic.

One way to achieve faster onboarding time is by decreasing the required cognitive load, and a strong modularization with independent repositories can help a lot with this. After all, if a new developer fixes a single bug, a smaller repository can be debugged and understood much faster than a large repository.

Also, the documentation can be kept up to date much more easily with a lot of smaller repositories than a few larger ones. Together with a smaller (and more focused) commit history, the entry-level is far smaller.

The downside is that complex bugs may be more problematic to debug. To understand a system fully, it's likely that a lot more expertise may be required. Knowing which code resides in which repository becomes a virtue and gives some senior developers and architects a special status that makes it difficult to remove them safely from a project.

A solution is to think in terms of feature teams that have main responsibilities but may share common infrastructure expertise.

Multiple teams

If we are capable of onboarding new developers quickly, we can also leverage multiple sources for recruiting. Recruiting great talent in **Information Technology** (IT) has become absurdly difficult. The top players already attract a fair share of all the top developers needed, especially in smaller ventures. How can new features be developed quickly when it takes months to acquire a single new developer?

A solution to this challenge is to consider the help of IT agencies or service providers who can offer on-demand development help. Historically, bringing in external staff can be a source of trouble too. **Non-disclosure agreements** (NDAs) need to be filled out; access needs to be granted. The biggest pain point may be procedural differences.

By using modularization as a key architecture feature, different teams may work in their own repositories. They can use their own processes and have their own release schedules. This can come in handy internally too but obviously offers the greatest benefit when working with external developers and teams.

One of the best advantages of being able to modularize the frontend is that we can come up with new ways to divide the teams. For instance, we can introduce true full-stack teams, where each team is responsible for one backend service and one frontend module. While this boundary may not always make sense, it can be very helpful in a number of scenarios. Giving the same team that made a backend API its frontend counterpart reduces alignment requirements and enables more stable releases.

Combining the ideas of **domain-driven design** (DDD) with multiple teams is an efficient way to reduce cognitive load and establish clear architectural boundaries.

DDD is a set of techniques to formalize how modularization can be performed without ending up in unmaintainable spaghetti code. In *Chapter 4*, we'll have a closer look at these techniques, which can help us to derive a modularization referred to as domain decomposition. However, deriving the right domain decomposition is not at all easy. Keeping the different functional domains separated is key here.

Isolated features

My favorite business driver for introducing micro frontends is the ability to ship independent features using their own schedules.

Let's say a **product owner (PO)** has an idea for a new feature. Currently, it is not really known how this feature will be accepted by users. What can be done here? Surely, we want to start with a small POC presented to the PO. If this is accepted, we can refine it and publish a **minimum viable product (MVP)**, which would serve to gather some end-user feedback.

Historically, a lot of alignment would be needed just to get this feature into the application in the desired way. Using micro frontends, we can roll out the feature progressively. We would first only make it available to the PO, and then adjust the rollout to the target end users too. Finally, we could release it in full regions or worldwide.

A selective rollout is only possible when a feature is isolated. If we have two features – feature A and feature B, with A requiring B – then a progressive rollout is never possible for feature B. It would always need to be a superset of the rollout rules of feature A.

In general, such dependencies may still be possible (or even desired), but they should never be direct in nature. As such, all available features may only weakly reference other features. We will see later that weak references form a basis for scalability and are crucial for sound micro frontend solutions.

A good test to verify whether a given micro frontend is well designed is to turn it off. Is the overall application still working? With this change, the application may not be useful anymore, but that is not the key point. If the application is still technically working, then the micro frontend solution is still sound. Always ask yourself, does it – in principle – work without this module?

Being able to turn off micro frontends consequently allows us to also swap individual micro frontends. As the inner workings do not need to be available, we can also fully replace them. This can be used to our advantage to gather useful user feedback.

A/B testing

We may find at least a dozen more good reasons to use micro frontends from a business perspective; however, one – unfortunately, often forgotten – reason is to simplify gathering user feedback. A great way to do this is by introducing A/B testing.

In A/B testing, we introduce two variants of a feature. One is variant A, while the other is labeled variant B. While A may be the current state of the feature (usually called the baseline), B may be a new way of approaching the problem. The following diagram illustrates this:

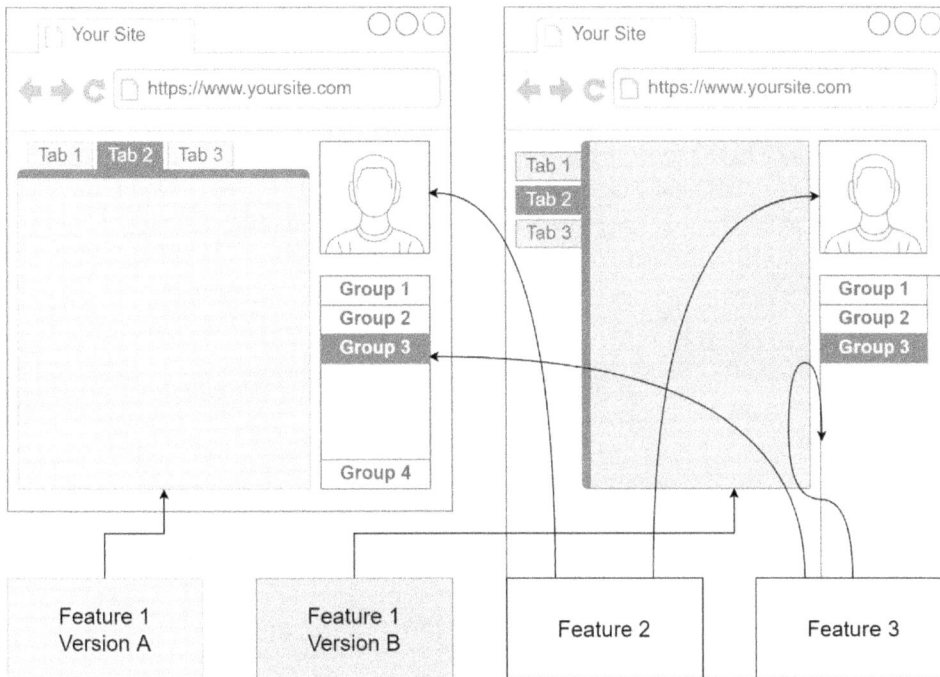

Figure 1.5 – A/B testing of a feature results in different experiences for different users

In scenarios where the feature is new anyway, an A/B test can be offered to all users interested in taking part. If we work with a baseline, it may make sense to survey only users who are new to the feature to avoid having existing habits contaminating the data.

Most frontend monoliths require quite a few code changes to include the ability of A/B testing for specific features. In the worst-case scenario, branching from variant A to variant B can be found in dozens – if not hundreds – of places. By introducing a suitable modularization, we can introduce A/B testing without any code changes. This is an ideal scenario, where only the part shipping the micro frontends needs to know about the rules of the active A/B test.

Actually, including the feedback and evaluating the results is then, of course, a completely different story. Here, micro frontends don't really help us. The connection to dedicated tools to support the data analysis is unchanged, with a monolithic solution.

Summary

In this chapter, we investigated how the web has evolved to reach a point where micro frontends can be presented as a viable solution. We discussed that the need for application scaling has hit a point where alignment efforts have had a massive impact on efficiency. Together with some new technical possibilities, the architectural pattern of micro frontends can now be implemented successfully.

You now know that micro frontends not only have a technical but also a business background. However, besides some advantages, they also come with disadvantages. Knowing from the start what to expect is crucial when aiming for a solid solution.

In the next chapter, we'll have a more detailed look at the challenges and pitfalls of micro frontends. Our goal is to be able to successfully tackle all challenges with sound solutions. We will cover some of the most difficult problems, following a path that leads to proper implementation.

2

Common Challenges
and Pitfalls

In the previous chapter, we saw that the evolution to microservices and ultimately micro frontends was more or less inevitable. The modularization of technology is a natural step that can be found in almost all industries. While modularization comes with a lot of strengths, it also comes with some challenges.

For our journey into micro frontends, we will also face these issues. In particular, we'll face all the common problems of modularization, plus additional trouble coming from our expectations about how a frontend should work.

This chapter covers the five most commonly seen issues when setting up a micro frontend solution.

We'll go into the following in depth:

- Performance
- Security
- Knowledge sharing
- Reliability
- User experience

The discussion on these will remain mostly theoretical. Specific implementation details will be discussed when we reach the different architectural styles.

Technical requirements

To follow the code samples in this book, you need knowledge of JavaScript and how to use the command line. You should have Node.js (version 20 or higher) installed using the instructions at `https://nodejs.org`.

For this chapter, there is no source code and no **Code in Action (CiA)** video available.

Performance

Performance is almost always regarded as a feature. While different kinds of applications may be satisfied with different levels of performance, especially for end-user-facing applications, the general mantra is *faster is better*. Amazon claims that every 100 ms of additional latency costs them about 1% in sales. This is a massive number and should not be underestimated.

Resource caching

As we saw in *Chapter 1*, there are multiple kinds of websites these days, such as the following:

- Static websites

- Dynamic websites (**server-side rendering** or **SSR**)

- Dynamic websites with AJAX

- **Single-page applications (SPAs)**

Every kind has its own techniques to boost performance and stay scalable. For instance, static websites will use an in-memory cache to avoid reading the page from the disk on every request. This simple feature can be already used in SPAs; however, it must be implemented by hand for SSR.

Most languages and frameworks that deal with SSR have no idea what kinds of dependencies are required to create a response for a given request. As such, introducing an automatic cache for these requests seems very difficult. We can assist these frameworks by giving hints.

There are multiple layers of caching that we can apply to help not only our server but also the browser to improve performance with a cache. Ideally, pretty much all frontend assets are uniquely named and can be cached indefinitely in the client.

> **Important note**
>
> There are several HTTP headers dedicated to properly communicating caching. While we focus on the strongest caching guarantees here, in general, more fine-grained approaches can be very useful, too. A full discussion on caching would certainly go beyond the scope of this book. See the following MDN link: `https://developer.mozilla.org/en-US/docs/Web/HTTP/Caching`.

At some point, you may be tempted to aggregate multiple assets into a single asset. For instance, if three micro frontends all produce a JavaScript file, you may want to bring this to a single JavaScript file. While this can be beneficial, the complexity and its outcome might not justify this investment. There are multiple reasons for this:

- HTTP2 streams requests and makes serving multiple resources almost as efficient as serving a single resource

- Caching becomes more complex and is less fine-grained, leading to more cache misses

- The individual scripts lose their context – for example, to automatically resolve URLs correctly

- One erroneous script can abort the execution of all scripts in the aggregation

Consequently, even consuming resources per micro frontend directly in the browser is usually better than a single large resource containing all micro frontends.

Bundle size

For SPAs or JavaScript-heavy pages in general, we need to be very cautious with the used JavaScript code. In today's landscape, it's far too easy to use multiple dependencies carelessly. Monitoring the resulting size of the JavaScript bundles and making sure that these bundles remain small is vital to great performance.

Especially for SPA-based micro frontends, we will use a bundler such as webpack or Vite to produce one or more JavaScript files for each micro frontend. Here, it is crucial to leverage bundle splitting to keep each JavaScript file rather small. A classic example would be to create a separate JavaScript file for each page in a SPA.

So, instead of writing code such as the following, you leverage the `import` function to introduce bundle splitting by lazy loading another JavaScript fragment:

```
import * as React from 'react';
import { Switch, Route } from 'react-router-dom';
import MyPage from './page';

export const Routes = (
  <Switch>
    <Route path="/my-page" component={MyPage} />
  </Switch>
);
```

In React, we can use the `lazy` function together with a `Suspense` component to allow lazy loading for parts of the application.

Updating the example to use lazy loading, we get the following:

```
import * as React from 'react';
import { Switch, Route } from 'react-router-dom';

const MyPage = React.lazy(() => import('./page'));

export const Routes = (
  <React.Suspense fallback={<div>Loading...</div>}>
    <Switch>
      <Route path="/my-page" component={MyPage} />
    </Switch>
  </React.Suspense>
);
```

For micro frontends, bundle splitting may be even more crucial than for non-micro frontend web applications. Since scalability is one of the core principles of most web applications, the only way to ensure a fluent user experience is to create smaller assets that are only loaded on demand.

> **Important note**
>
> It always makes sense to be aware of what dependencies contribute to the final bundle size. For webpack, the `webpack-bundle-analyzer` package visualizes this quite nicely. Also, websites such as `https://bundlephobia.com` help to identify the cost of including dependencies quite early.

In general, we should try to introduce the lazy loading of individual code as much as possible. As seen in the previous example, in SPAs, this can be as easy as lazy loading every component representing a page. Also, dependencies that are only used in "exotic" scenarios should not be part of the main bundles. Using modern `Promise` infrastructures helps a lot here.

Request optimizations

Micro frontends are all about modularization. As a result, the number of HTTP requests will definitely grow. After all, instead of having everything integrated into a monolith, we now have a system where the different technical parts are still very much separated – either on a server or in the browser.

Quite often, many modules will need to request the same or similar data. If we are not careful, this may lead to a lot of stress on the API servers. Even worse, aggregating the results will take longer, leading to a worse user experience. It can, therefore, be useful to mitigate this by introducing micro-caching for API requests. If a request to `/api/user` is cached for 1 s, then other execution units get the result immediately from the cache. Importantly, on the server side, this cache has to be exclusive to the page's request context.

Another factor to think about is request batching. Especially with a microservice backend, there may be many requests needed to get some required information aggregated. If we introduce a technology such as **GraphQL** on the API gateway, we could run a single request with multiple queries. The backend then resolves all the different queries – using a much faster and more direct way than the client could. Here, the backend may cache the queries, too.

Now that we've touched upon request optimization, with its potential security challenges, it's time to look a little into the security sector in more detail.

Security

Before we even start talking about potential vulnerabilities, we need to examine closely that there are two extreme standpoints regarding security when talking about micro frontends:

- Centrally provided cross-cutting concerns including user management
- Full isolation where each module is responsible for its own user management

Both extremes come with their own advantages and disadvantages. As already seen, finding the right balance is key to a successful implementation. There is no silver bullet – a lot depends on the boundary conditions set by the project.

By far the biggest challenge regarding security management can be found in the execution of potentially untrusted code. Since code is no longer centrally created and maintained, the likelihood of unwanted snippets entering the system increases drastically. Consequently, further boundaries and processes are required to avoid vulnerabilities.

Let's dissect the two options for user management, as well as the challenge of script execution in more detail.

Central user management

This makes sense when there is a single backend that should be used. For many problems, this extreme will be close to the ideal solution.

In central user management, we either run all HTTP requests to the backend via a central function or use cookies to implicitly append the session to requests targeting the backend. In the former case, a token may be appended to requests created by the central function. In the latter case, tokens should not be available in the application at all. Here, user information may be stored centrally to simplify answering common questions regarding the current user.

Figure 2.1 shows that the micro frontends need to go through the user management module of the application to communicate with the backend. This way, the token gets inserted properly. The user management module is rolled out with the foundation layer of the micro frontends solution and is therefore independent of the individual micro frontends:

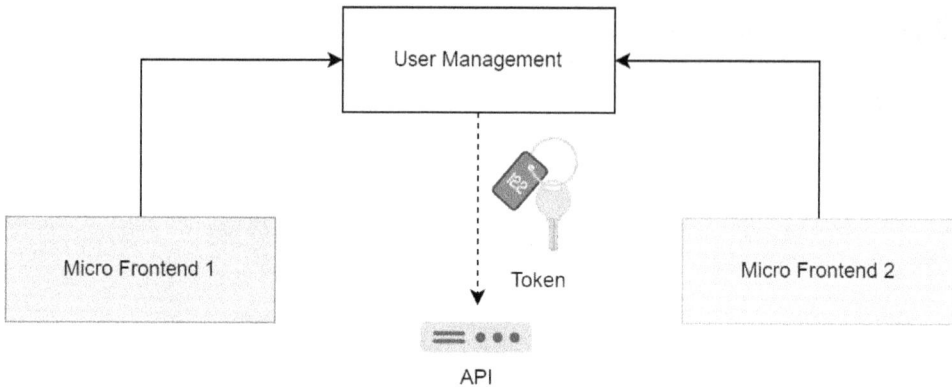

Figure 2.1 – Central user management requires each backend call to go through a central instance

Through middleware, the user management part could be decoupled from the API requester logic. This will use HTTP header injection or other techniques depending on the means of transportation.

Individual user management

Individual user management makes sense when there are multiple backends or micro frontends coming in from third-party sources.

Here, each micro frontend holds, for example, a token for backend communication in memory. This plays quite well with the micro-caching of requests to avoid having to regenerate dozens of tokens for the same application in the backend. *Figure 2.2* is a visual representation of this:

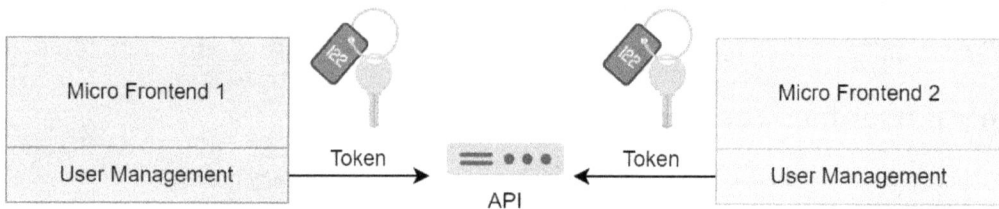

Figure 2.2 – Individual user management potentially allows calling APIs directly

The biggest advantage of individual user management is the guaranteed independence of individual micro frontends. There are no changes to worry about, which may break one individual module.

While the decision between central and individual user management is important, an even more crucial area with respect to security is the actual script execution.

Script execution

No matter whether we run our micro frontends on the client, the server, or both, we need to be aware that code coming from one micro frontend might have an impact on other micro frontends or the application as a whole. As such, one of the first things to clarify is who will be able to contribute micro frontends to the application. If the answer is "anyone," then we need to think about quite strong isolation. If the answer is "only internal teams," then we don't have to implement anything special. Still, having a reliable system makes sense.

To achieve great isolation, various efforts are necessary. For instance, in the browser, we may place the scripts inside an <iframe> instance or in a new Worker instance, just like so:

```
// create worker
const myWorker = new Worker('./my-worker.js');

// receive messages from the worker
myWorker.onmessage = e => {
  console.log('Received new data', e.data);
};

// message the worker using any object that can be cloned
myWorker.postMessage({ type: 'hello', text: 'Hi World!' });
```

In Node.js, we could fall back to creating a new vm.Script instance:

```
// import vm module
const vm = require('vm');

// create context with no globals except require and
  // console
const context = vm.createContext({ require, console });

// create a new script execution environment and run it
const script = new vm.Script('require("./script.js")');
script.runInContext(context);
```

In any case, the communication with the host and the available APIs should be strongly regulated. This will require quite some balance between flexibility and security.

As an example, there are two ways to do central user management in the browser: via an implicit cookie from the backend or by storing some token. While the former requires a **Backend For Frontend (BFF)**, the latter requires some secure token format – for example, using signed **JavaScript Web Tokens (JWTs)**. Depending on how these are obtained and what security properties they contain (for example, expiration time), these can be regarded as secure for practical purposes.

Since cookies work pretty much out of the box and can be secured implicitly by the browser, we can almost always use them. One of the browser's security guarantees is that an `HttpOnly` cookie will never be read out by a script. As such, the script execution – even of scripts that have been authored by externals – can be considered sufficiently secure in that regard.

Scripts coming from externals should still be sandboxed as much as possible. Without prior static code analysis or manual review, they should not be allowed to be part of the execution environment. Otherwise, bringing in – for example – a keylogger that reports back to a malicious source is way too easy.

Another challenge to crack is implicit knowledge sharing, which is usually quite hidden in distributed applications.

Knowledge sharing

Ideally, each micro frontend represents an isolated module, which works without dependencies and without any knowledge of the other micro frontends. Realistically, micro frontends will have dependencies and at least some knowledge of other micro frontends. These dependencies are usually the result of referencing parts such as components of other micro frontends.

There are two kinds of references that can be created from a micro frontend:

- Direct (or strong) references leading to tightly coupled modules
- Indirect (or weak) references leading to loosely coupled modules

Only loosely coupled modules can scale well. The problem with loosely coupled modules is that some conventions and contracts still need to be followed. For instance, if we emit an event with a certain name, potential listeners expect this name to remain the same. Once we change the name, the listeners cannot receive the associated event anymore – they'd need to change that name, too.

The agreement for an identifier is what we refer to as knowledge sharing. There are multiple ways to perform the act of knowledge sharing:

- Two or more teams agree on a certain name
- One team picks the name

Instead of an identifier, the agreement could also be formed on certain behaviors, flows, or technology decisions such as shared dependencies and their versions.

Figure 2.3 shows one example where the team that created the event has to communicate to all-consuming teams that some change has happened:

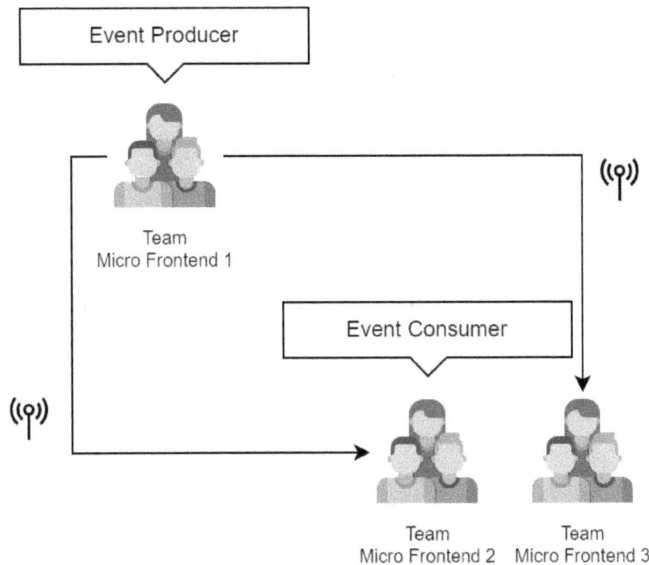

Figure 2.3 – Every change to the event structure requires alignment with interested parties

Effectively, any kind of agreement that is used in the micro frontends implicitly forms a contract between them. Since this contract is not enforced by some tooling (for example, a compiler), it is a potential source of error. How can we enforce this?

Here, DevOps best practices come into play. After all, micro frontends do not enter our application on their own. They must be approved or go through some predefined build pipeline. In any case, we will have a central gate somewhere to control what is entering. This gate can be used to technically enforce conventions, check relationships, and perform integration tests.

If a breaking change is necessary, then the principles of change management come in handy. This is not only true with internal micro frontends but is still applicable with external micro frontends. The communication factor plays an important role.

> **Important note**
> While the reduction of alignment is one of the goals of micro frontends, it can never be fully removed. The need for communication and knowledge sharing will always exist. The good news is that the principles and ideas for efficient developer communication are quite universal and still apply to micro frontends. For a strategy to communicate API changes, see `https://freecontent.manning.com/designing-apis-communicating-with-your-developers/`.

Having a breaking change without any grace period should be avoided at all costs. One preference is to start with a deprecation, then a stub, and then to remove the relevant part. This can also be included in the change management process:

- A deprecation should yield a warning to developers at compile time

- A stub could report back to the owner at runtime

- The removal will break this feature for the end users

Integration environments make this process even less frightening by gaining useful insights before reaching production with end-user visibility. They are also wonderful for verifying the reliability of an application.

Reliability

Micro frontends load dozens of smaller modules from various sources. Every request has a certain chance of failing. As a result, the likelihood of some modules failing to load is relatively high and should not be underestimated.

A good test to verify that the given application is, architecture-wise, on solid ground is to just disable one of its modules. Does it fail? Was the failure expected? Is an error reported? All these questions can then be answered. Indeed, while disabling the login functionality will result in a pretty much unusable application, the application itself should just keep on working. If the application is not working anymore, we've introduced too much coupling.

As with scalability, the best way to approach the reliability topic is by introducing loose coupling. Furthermore, the standard techniques to make HTTP requests more robust can be used, too. Here are the techniques:

- Set (very) small timeouts

- Introduce retries for critical requests

- Detect a connection loss and handle it gracefully

Still, a single micro frontend will crash. In such a scenario, the most important point is to place an error boundary exactly where the micro frontend started. At all costs, the main application, including all the other modules, should be kept alive.

Having a reliable application forms the basis of a great user experience. For micro frontends, however, we'll need to work a bit harder to get a consistent user experience.

User experience

One of the key differences between microservices and micro frontends is that the latter needs to appear as a single unit to the end user. This requirement impacts not only the visual design but also behavior and wording. If every part of the frontend is independently developed and maintained, how can we ensure that a single coherent application is presented to the user?

Wording

If we think of the micro frontends as molecules consisting of atoms, we could try to harmonize the situation by aligning available atoms. Here, atoms would be represented as components such as a form, a card, or a text field. Standardizing these would certainly help but would still leave out a design vision or specific wording.

As it turns out, there is no silver bullet here. In the end, it's a trade-off between centralization and independence. Where we define the point for our application is up to us. For wording, one option is to do the following:

- Put a set of shared wording fragments in a central wording container
- Allow a dedicated wording team to refine and overwrite fragments later
- Leave specialization and specifics to each micro frontend

This represents a suitable compromise for many applications. It unblocks developers to ship features fast while allowing a central team to align and adjust the wording later on.

For the design vision, a style guide makes sense. Usually, a style guide is centered on a pattern library.

Pattern libraries

Defining a set of components to be used across the whole application is called a pattern library. While the term *library* is most often referred to as a technical dependency for our application, it can also just mean *available CSS styles*. Even just *styling specified on paper* is an interpretation of a pattern library.

Independent of the specific interpretation, there is a single standard to transport components, including their designs and behavior.

The challenges of designing components that are flexible enough to be used in micro frontends will be further discussed in *Chapter 18*.

Summary

In this chapter, you learned what core challenges exist when dealing with micro frontends. You saw how little tricks can improve performance drastically. With additional awareness in the security sector, you can now build robust solutions that cannot be easily attacked.

Being efficient in knowledge sharing within development teams is as important as having a consistent user experience. Both areas require discipline and strict guidelines. Using a centralized but distributed approach is quite often a good compromise to create enough context without putting too many constraints on different modules.

In the next chapter, we'll cover different deployment scenarios to approach how we can structure and implement micro frontends tailored to solve a specific problem.

3
Deployment Scenarios

In the previous chapter, we've looked at common challenges and pitfalls when applying micro frontends. Even though those issues make our lives a bit more complicated, it's quite often a necessary evil to get the desired benefits: improved scalability in terms of development teams. The area in the software development life cycle where this benefit can be best seen is deployment.

Monoliths come with a single way of deploying: all or nothing. This means that the full business logic is rolled out in a single activity. The reason why monoliths scale worse in development than their micro frontend counterparts is that they come with these larger releases. Very often, large releases result in longer release cycles. After all, testing and verification need to happen too.

For micro frontends, we have a lot of options for creating deployment pipelines. Even here, we can leverage a single pipeline. Alternatively, we can also use the other extreme of the spectrum: having distributed deployments with many independent pipelines. Finally, we can also aim for a solution that uses aspects of both: a central and a distributed deployment scheme.

This chapter covers these three scenarios, with some potential variations. For illustration purposes, we'll use **YAML Ain't Markup Language** (**YAML**) files, as found in **Azure DevOps** and **CircleCI**, though the ideas can be transported to most **continuous integration** (**CI**) systems. We will not assume a specific micro frontend architecture.

This chapter will cover the following main topics:

- Central deployments
- Distributed deployments
- Hybrid solutions

Technical requirements

To follow the code samples in this book, you need knowledge of JavaScript and how to use the command line. You should have **Node.js** (version 20 or higher) installed using the instructions at `https://nodejs.org`.

For this chapter, there is no source code and no **Code in Action (CiA)** video available.

Central deployments

While having one large release is often considered a downside of monoliths, it also can be seen as a desired solution. After all, this is an easy way to ensure we have one unit that works together. In this solution, we can also control when the application updates very well. Finally, the ultimate advantage of using a central deployment for micro frontends is that we can join the micro frontends that are already up front. This allows optimizations and enhancements that would be very hard to incorporate in a non-central pipeline.

It turns out there are two major ways of thinking about a central CI pipeline, outlined as follows:

- Using a **mono repository (monorepo)**
- Joining multiple repositories

While the former is easier to set up, the latter may be closer to what the majority of users are after. Let's have a look at both.

Using a monorepo

Monorepos have become quite popular, one reason being that they allow the mixing of different packages without having to wait for deployments. That way, we can get a working state of different packages without having to go through many deployment cycles.

A monorepo is naturally qualified for micro frontends but is also questionable. After all, if we have all the information in one place, why do we need to scatter it out for consumption? In the end, this resolves well if we think about micro frontends as another metric for development without having the need for deployment independence.

Figure 3.1 illustrates how monorepos can be used with a single central CI pipeline. Every micro frontend in the monorepo is built and published within the same pipeline.

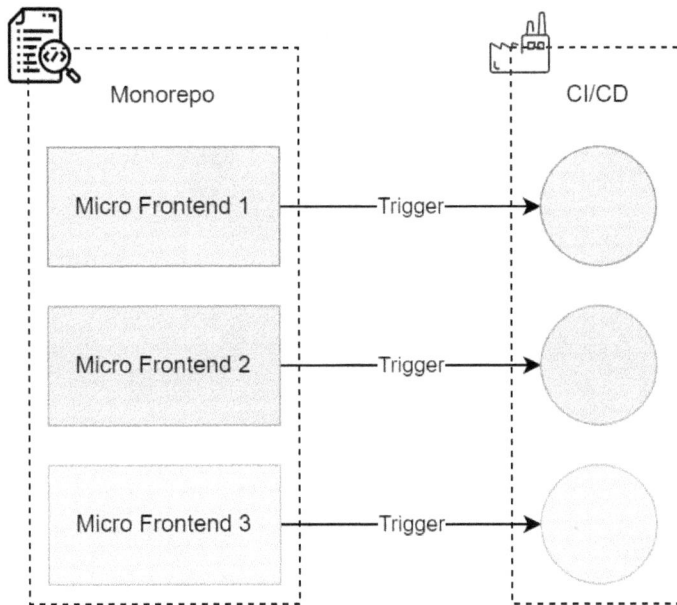

Figure 3.1 – In a monorepo, any code change triggers a joint build of all micro frontends

In frontend projects, monorepos are usually handled by tools such as **Lerna** or **Nx**. These tools know what a monorepo is and how to connect the different packages within the monorepo correctly. They can be thought of as a companion for package managers such as **npm** providing an improved developer experience for monorepos. For this book, we use Lerna version 8.

> **Important note**
>
> Lerna is a task runner based on Nx that makes working with monorepos more convenient. Together with an efficient package manager such as **Yarn**, it can resolve dependencies while referencing the contained packages correctly. The most important commands are `lerna run` and `lerna publish`. It is capable of doing unified versioning, where every contained package gets the same version, or independent versioning, where every contained package decides on its own version. More information can be found at `https://lerna.js.org/`.

Once we have initialized a new Node.js project using `npm init -y`, we can run `npx lerna init --packages="packages/*"` to integrate Lerna properly. This will set up a *packages* folder where the individual micro frontends – or other libraries – would be developed.

Adding a CI pipeline using Azure DevOps to the monorepo can be as straightforward as adding an `azure-pipelines.yml` file to the root of the repository next to the `package.json` file. The content can be as simple as this:

```
trigger:
- release

steps:
- task: NodeTool@0
  inputs:
    versionSpec: '20.x'
    displayName: 'Install Node.js 20'
- bash: |
    yarn install
  displayName: Bootstrap repository with Yarn
- bash: |
    npx lerna run test
  displayName: Run rests in all packages
- bash: |
    npx lerna run build
  displayName: Run build in all packages
```

For CircleCI, which is another cloud service to host CI/CD pipelines, the setup is similar. The main difference is that we use a file called `config.yml` instead, which has to be placed in the `.circleci` directory. This file contains the following code:

```
version: 2
workflows:
  version: 2
  default:
    jobs:
      - build:
          filters:
            branches:
              only: release
jobs:
  build:
    docker:
      - image: circleci/node:20
working_directory: ~/repo
    steps:
      - checkout
      - run:
          command: yarn install
```

```
        name: Bootstrap repository with Yarn
  - run:
      command: npx lerna run test
      name: Run rests in all packages
  - run:
      command: npx lerna run build
      name: Run build in all packages
```

This is sufficient to get a repository properly set up, tested, and built. For validation purposes (for example, in a pull request), this is wonderful, but for actually deploying, it is not sufficient. Here, we can create different stages to distinguish between a build phase and a release phase properly.

Joining multiple repositories

A monorepo is a great way to get multiple artifacts deployed without much need for maintenance. However, to enable great development scalability, we may need to introduce multiple repositories. This does not necessarily mean that each repository needs to set up a dedicated pipeline. Independent repositories will trigger the central pipeline, combining all the code into a single solution.

Figure 3.2 illustrates the path that micro frontends from separate repositories take to come together in a single pipeline. Ultimately, no matter where the code changes, it's the same pipeline that triggers, combining all repositories in a single build:

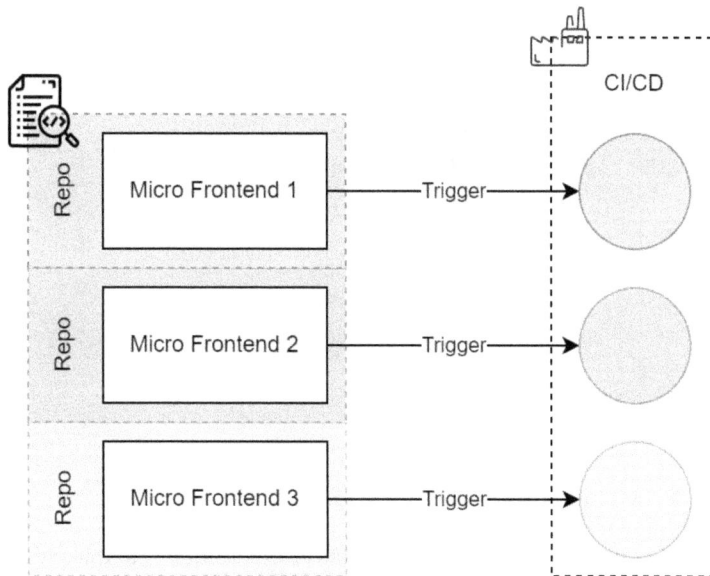

Figure 3.2 – Multiple micro frontends coming together in a single
pipeline, no matter which repository changed

The beauty of this solution is that development – and even maintenance – is apparently independent, but a single coherent solution may still be produced by the CI pipeline. Even if we use the micro frontends independently later, we can use the central pipeline to obtain crucial information or make checks against the solution as a whole.

The downside of this approach is that this is still a central pipeline, meaning that all knowledge is stored centrally. If a new repository is added to the solution, the pipeline needs to be updated to be aware of the repository.

A potential approach to a working pipeline definition in Azure DevOps looks like this:

```
resources:
  repositories:
  - repository: MicroFrontend1
    type: github
    endpoint: GitHubServiceConnection
    name: ArtOfMicroFrontends/MicroFrontend1
    trigger:
      - release
  - repository: MicroFrontend2
    type: github
    endpoint: GitHubServiceConnection
    name: ArtOfMicroFrontends/MicroFrontend2
    trigger:
      - release
  - repository: MicroFrontend3
    type: github
    endpoint: GitHubServiceConnection
    name: ArtOfMicroFrontends/MicroFrontend3
    trigger:
      - release

steps:
- checkout: self
- checkout: MicroFrontend1
- checkout: MicroFrontend2
- checkout: MicroFrontend3

- script: dir $(Build.SourcesDirectory)
```

Here, the pipeline itself would come with a set of scripts or other utilities to make the CI work. The different repositories can still be adjusted individually – for example, to trigger on different branches or conditions. Nevertheless, these adjustments need to be done centrally, which is why individual teams would not be fully empowered.

> **Multiple repositories with CircleCI**
>
> Out of the box, this feature is not provided. A workaround would be to explicitly clone additional repositories via `git clone` and work with their sources in the central pipeline. To trigger the central pipeline, individual pipelines would be required that use the **HyperText Transfer Protocol (HTTP)** **application programming interface (API)** from CircleCI to start the process.

To empower individual teams, distributed deployments come in handy.

Distributed deployments

The other extreme in terms of deployment is to create a fully distributed deployment system. Here, each micro frontend needs to come up with its own CI pipeline, which could leverage any kind of provider outside. Information regarding which micro frontends are available would be aggregated either by a central service or some federated approach, where individual micro frontend mirrors are known in a registry. This registry may be distributed, too. In practical scenarios, such a registry would be centralized.

A big advantage of distributed deployments lies in the independence of the individual micro frontends, whereby each team is fully responsible for their artifact. This statement is not only written on paper but truly exists in the basic setup of the application itself. If a team is unable to set up a proper release pipeline, nothing will happen and their contribution will never be noticed.

As with central deployments, we can look into two different scenarios here, too: either using a monorepo as a code base or starting with individual repositories to have dedicated pipelines.

Using a monorepo

The monorepo case starts in a similar way to the central deployment scenario: we place all relevant micro frontends in a single code base. The difference now is that each team is allowed to host their own CI pipeline for their micro frontends living in the monorepo. For example, we could have three different pipelines for three micro frontends developed in the monorepo.

Figure 3.3 shows the relation between the individual pipelines and the monorepo. A single code change in the monorepo would trigger all pipelines.

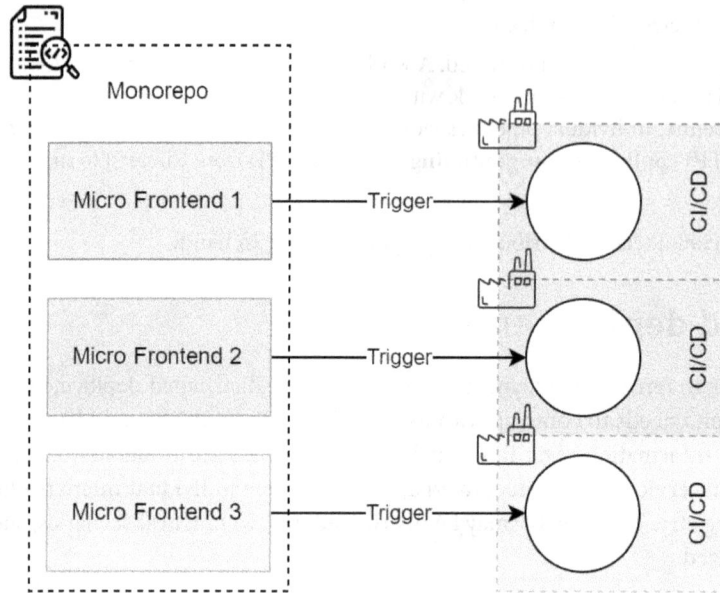

Figure 3.3 – A single code change triggers multiple pipelines that may or may not continue

Quite often, the first thing a pipeline would do, when triggered, is a comparison, in order to find out whether anything changed in the relevant code sections. Without some change detection, all pipelines would trigger a release – even though there is nothing new to release.

While running a pipeline unnecessarily sounds harmless at first, it presents a problem in terms of economics (running CI pipelines longer than necessary represents an unnecessary cost), user experience (through potential cache invalidation, the user gets longer wait times before the application is started), and security (every deployment presents an opportunity to sideload code from an attacker, even though these vulnerabilities are rare and unlikely).

The simplest way to do change detection is by introducing individual versioning in Lerna. This could be done when initializing the monorepo using the `npx lerna init --independent --packages="packages/*"` command. Alternatively, change the `lerna.json` file to set the `version` property to `independent`, as illustrated in the following code snippet:

```
{
    "npmClient": "yarn",
    "version": "independent"
}
```

Having independent versioning, we can just compare the version set for the micro frontend that we care about against the last released version of it.

An alternative is to either hash the sources or compare a hash value of the build artifacts. The former is a bit more unreliable, while the latter is definitely more expansive as we need to run a full build before being able to answer the question.

A potential way out of this is to split the monorepo into multiple repositories.

Using dedicated pipelines

Micro frontends start to shine fully when we use one repository per micro frontend. This way, we get as much freedom as possible. However, as a downside, all sorts of different decisions (for example, which testing framework to use, which practices to follow, which naming conventions to consider, and so on) need to be made, too.

Figure 3.4 illustrates the relationship between code repositories and CI pipelines in this scenario. Here, only a code change in the respective micro frontend repository triggers the start of its pipeline:

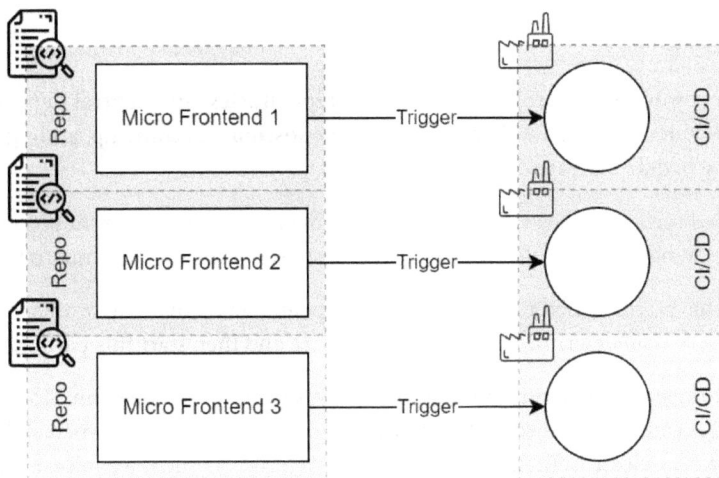

Figure 3.4 – Each repository has its own pipeline and is therefore completely independent

When we start using independent pipelines to deploy our different micro frontends, we also lose the ability to perform any build-time steps to optimize or recombine the solution. Surely, we could integrate this on some server, but this would require additional logic.

Therefore, in this scenario, the micro frontends are usually combined dynamically at render time, which is what most people actually regard as the baseline definition of micro frontends. As we will see later on, this can be on a server (see *Chapter 7*) or within the browser (see *Chapter 9*).

If we want to keep some optimizations at build time, we should try to combine a central deployment approach with a distributed solution. The result is a kind of hybrid approach.

Hybrid solutions

Quite often, the main reason for introducing micro frontends is the scalability in terms of development teams. Therefore, as outlined, going in the direction of distributed deployments seems natural and has the biggest scaling impact. Nevertheless, for one reason or another, there may be a collective of micro frontends that should be deployed – or at least developed – in a single repository. Now is the time to talk about a third way of doing deployments using micro frontends – namely, a hybrid solution.

Hybrid solutions try to leverage the advantages of both extremes: a central deployment solution and a distributed deployment system. As a result, challenges that come with hybrid solutions are usually a combination of challenges pertaining to both solutions.

Usually, going for a hybrid solution is motivated by an ability to have scheduled releases for a set of micro frontends. An alternative is to just trigger a release when any of the micro frontends change. We'll have a look at both ways and their potential areas of use.

Scheduled releases

Logically, we can combine monorepos with normal repositories, and central deployments with independent deployments. Once we do that, many new possibilities come up, among which is the opportunity for a scheduled release.

Having a scheduled release makes sense if we want to make sure that different pieces fit together just perfectly. We can have a scheduled release just for a subset of the available micro frontends, too.

The easiest way to have a scheduled release is to have a pipeline set up that only works when triggered manually. This way, we can determine a date for our release and then start this particular pipeline.

An alternative is to release frequently and automatically, such as every week on Monday morning. Having a time trigger can be combined quite nicely with a certain branch setup, where only a particular branch (for example, a release branch) would be used as a basis. As such, we release frequently, but only the parts of the code that have been properly selected for the next release.

Another alternative is to trigger a release on change.

Triggering on change

Triggering the start of a joint release pipeline on change can be appealing, especially if we use a dedicated branch to mark releases anyway. Usually, we would not consider joining multiple repositories into one big pipeline here, which means that only a subset of micro frontends living in the monorepo is considered in the large release.

As shown in *Figure 3.5*, the split between the core micro frontends coming from the monorepo and the auxiliary ones coming from the other repositories is also reflected in the CI setup. Here, we can leverage a distributed form of development while keeping cross-cutting concerns nicely packaged in a single coherent repository:

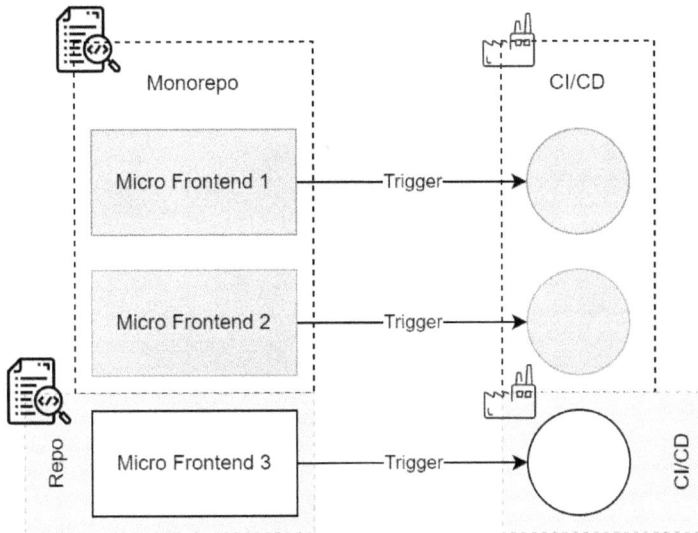

Figure 3.5 – Some micro frontends release independently, while others have a joint release

One example where mixing central deployments with distributed deployments makes sense is to perform optimizations regarding used dependencies. Having information on some micro frontends available at build time allows us to introduce a dependency analysis, uncovering new (or unused) shared dependencies.

Summary

In this chapter, you learned about the different deployment scenarios that are possible with micro frontends. We introduced a central deployment scheme to illustrate how a monolithic release cycle can be applied to micro frontends. Using distributed deployments, we are able to make every team truly independent.

Quite often, we will take a middle way here. For this, you saw one example covered by the hybrid solution. Here, we can control all the cross-cutting concerns and, most importantly, micro frontends, while still giving each team the independence they need.

In the next chapter, we'll cover the topic of **domain decomposition**, which helps us to define which micro frontends should exist. Domain decomposition also essentially helps us to find the appropriate team splits.

Domain Decomposition

In the previous chapter, we saw that micro frontends offer a broad variety of deployment options. An obvious question is: *how do we decide which model to use?* The answer to this question lies somewhere between the technical requirements of the project and the team setup. To leverage the team setup in the best possible way, we also need to choose a proper strategy for identifying the scope of the individual micro frontends. This is where the technique of domain decomposition helps.

Domain decomposition allows us to come up with a consistent way of breaking up our problem domain into smaller pieces that can be developed and deployed independently. This may sound simple at first but turns out to be quite complex. Even though the essentials of this chapter are not very technical (or practical) in nature, they should be carefully studied and applied before any actual implementation work. After all, the domain decomposition strategy lays the foundation of the architecture of our solution.

To grasp the concept fully, we first start by introducing the principles of **domain-driven design** (**DDD**). Here, we follow the thinking of Eric Evans in his famous book *Domain-Driven Design: Tackling Complexity in the Heart of Software*. Afterward, we will discuss the importance of **separation of concerns** (**SoC**), and we'll see that there are two essential decomposition strategies: a technical strategy and a business-driven strategy. Finally, we'll look at the architectural boundaries that need to be defined before any implementation as well.

Here are the main topics that we will be looking at:

- Principles of DDD
- Separation of concerns
- Architectural boundaries

Technical requirements

To follow the code samples in this book, you need knowledge of JavaScript and how to use the command line. You should have Node.js (version 20 or higher) installed using the instructions at `https://nodejs.org`.

For this chapter, there is no source code and no **Code in Action (CiA)** video available.

Principles of DDD

The idea of DDD was first popularized by Eric Evans. In his book, he describes the basic pillars that form DDD, according to his vision. While the book certainly has a lot of truth to it, the entire army of ideas was probably rarely—if ever—realized in real-world projects. Let's try to distill the most important ideas for our use in micro frontends.

When we refer to DDD in the context of micro frontends, we will not include parts such as value objects or the need for a ubiquitous language. Instead, we almost exclusively take DDD as a blueprint to help us with the following two things:

- Defining micro frontends with clear boundaries
- Coming up with a strategy to establish these boundaries

While DDD uses the word *modules*, which was back then an alias for *packages*, we will refer to this unit as a *micro frontend*. The other thing that DDD introduces is a so-called bounded context. Finally, DDD defines a context map to make sense of it. But let's back off and look at each area separately.

Modules

A **module** serves as a container for a specific set of functionalities in your application. For a web application, this could relate to a full page or parts of a page. For instance, a module dealing with the order process could be responsible for showing a cart symbol on the pages of a web shop, such as the product catalog or product details.

As mentioned, a module in DDD is usually what we will bring to the table in the form of a micro frontend. Nevertheless, keeping to a general notation here makes sense; after all, this helps us to see that good architectures rarely have anything to do with specific implementations but actually would work in multiple variants. In micro frontends, the functionality will be mostly designed around components.

A module is, however, a part of the full problem domain, focused on a specific area (or subdomain). The general design principles for modules are low coupling and high cohesion. As such, a micro frontend should represent a single unit to solve one problem without relying on any other micro frontend directly.

In reality, we will be tempted to rely on other micro frontends to simplify the code. The one strong advice here is to avoid doing that, as the immediate simplification will usually backfire in a more complicated and less flexible solution later. When we start looking at the different architectural patterns later in *Chapter 7* onward, we'll see how to decouple the micro frontends.

To help us figure out what could be part of a module, DDD introduces the theoretical concept of a bounded context.

Bounded context

A **bounded context** is used to define the boundaries of the functionality of a subdomain. This is an area where only the functionality from a certain domain makes sense.

While this may initially seem similar to a module, it is actually quite different. Multiple modules live within the same bounded context. The bounded context is an umbrella connecting them and has nothing to do with runtime considerations.

Bounded contexts are also broader than modules as they contain both unrelated concepts and shared concepts. Naturally, this will result in some overlap and duplication between different bounded contexts.

The following diagram illustrates the relationship of two bounded contexts within an example problem domain:

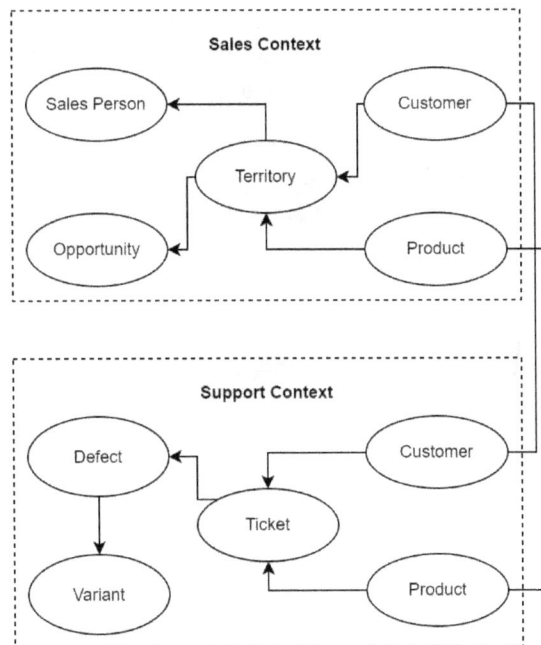

Figure 4.1 – Relation of two bounded contexts in an example problem domain

How we split a bounded context into modules is up to us. In the end, however, a more critical question is, *what are the bounded contexts in our problem domain?* There is no blueprint here, but a nice way to visualize and think about contexts generally is in the form of a context map.

Context map

A **context map** is a useful tool to illustrate the involved contexts of a system, including their connections. Instead of reusing objects from other contexts, a transformation defined by the context map should be used to always create domain-specific objects.

For our design decisions, a context map can be helpful to see where sharing occurs and how to still produce isolated micro frontends. To do this, we need to first identify these self-contained domains. This is where two principal ways come into play: strategic domain design and—as an alternative— tactical design. Let's look at both.

Strategic domain design versus tactical design

While strategic domain design helps us to extend knowledge of a problem domain and come up with useful guidelines, tactical design tries to come up with design patterns and building blocks that form a system. In other words, strategic domain design identifies the different domains and the communication between them, while tactical design is all about structuring these domains.

In our micro frontend space, we have the choice of leaving the tactical design to the implementers of the micro frontends, giving them some guidance, or predefining a structure that needs to be followed already. We will see later on how these architectural boundaries can actually be decided.

In contrast, strategic domain design is helpful for decomposing a full problem domain into smaller (sub) domains. Starting with a full problem domain, we derive the individual subdomains and extract their bounded contexts. The relationship between the contexts is then captured in a context map. The following diagram shows this theoretical process:

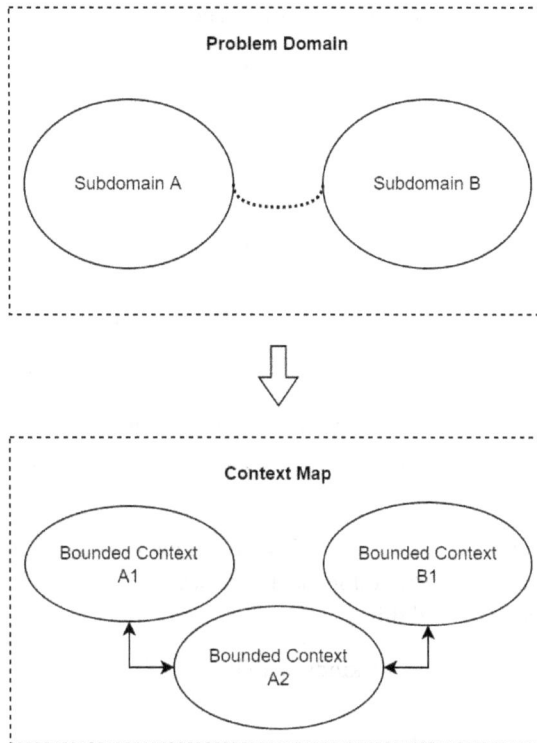

Figure 4.2 – The idealized process of deriving a full domain decomposition
from a given problem domain, including a working context map

Let's try to understand the process with an example. We have a library that wants to go online. They want to allow lenders to monitor their status, and visitors should be able to browse a catalog. Here, we can identify the following subdomains:

- Lender information
- Checkout process
- Book details
- Book catalog

That's about it. Surely, we could add some of the more generic technical domains such as user management—or authentication and authorization—to the preceding list; however, initially focusing on the problem alone makes sense.

From these four subdomains, we can derive the following bounded contexts:

- Lender user information

- Lender book-holding information

- Checkout process

- Public book details

- Lender book details

- Public book catalog

- Lender book catalog

The lender book details connect potential lender information with the book details to inform an authenticated user about how long a book could be lent out. Likewise, the lender book catalog tries to make this information available on a catalog level.

Not every one of these bounded contexts needs to become its own micro frontend, but it surely does not make much sense to go beyond them immediately and without good reason. Splitting a problem domain too drastically is hardly a good thing.

For now, we need to back off and see which kinds of splits actually make sense.

Separation of concerns

In splitting discussions, we'll rather quickly end up in a situation where it's more about the kind of split rather than the domain to split. Interestingly enough, when micro frontends started, most people were excited about technical splits. In my opinion, however, the true benefit of micro frontends can only be realized with business-driven split decisions.

Let's have a look at both options, what they have in common, and where they excel.

Technical split

A technical split usually starts by looking at the screen and drawing lines on the web page. Quite naturally, we will end up with a split that may be close to the illustration shown in the following screenshot:

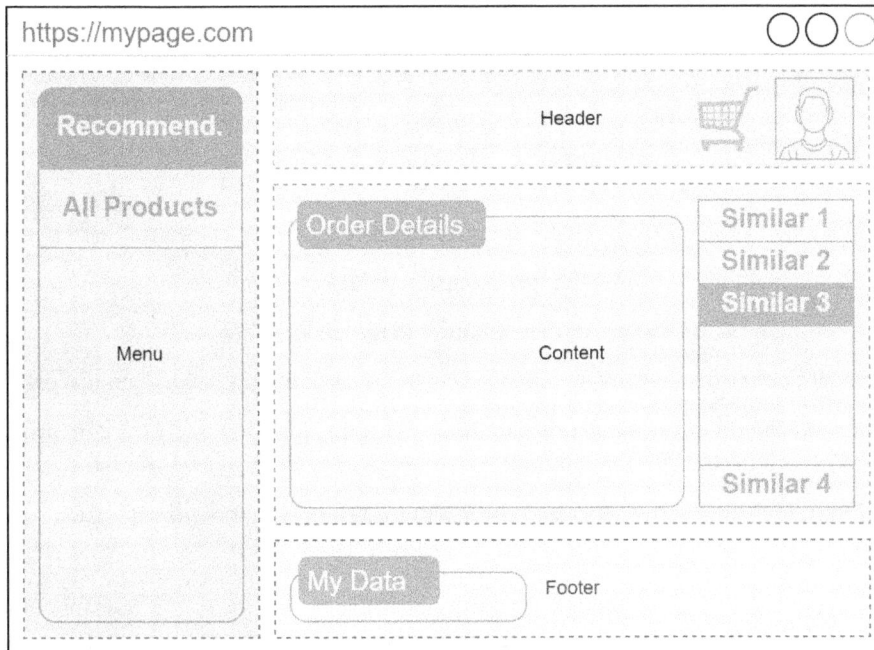

Figure 4.3 – A possible technical split; elements on the screen are
grouped and placed in different micro frontends

The problem with a technical split is that each micro frontend now contains parts of different domains—for instance, the navigation micro frontend as shown in *Figure 4.3* will contain links to user management, as well as some ordering functionality. While this seems okay at first, it presents a problem in the long run. After all, how does the navigation micro frontend know about all these different parts?

A useful technique to find problems with such architectural decisions is to trace how localized changes are. For instance, if we introduce a new micro frontend (for example, a billing micro frontend) do we need to touch other micro frontends too? If yes, how many? If the introduction of a new micro frontend involves changes in all micro frontends, then something is not right. This is a clear indicator that our architecture is not working out.

Ideally, a new micro frontend is completely self-isolated; therefore, no changes are required to any other micro frontend. This is one of the advantages of using a functional split.

Functional split

When we split an application into its functional parts, we see immediately that one screen almost never belongs to only a single micro frontend. After all, this way, there would rarely if ever be any connection from one data source to another. Likewise, all the interactions would be rather static and not very efficient.

The challenge with a functional split is that it is quite abstract and rather tedious to implement. The main reason for the implementation difficulties is that everything needs to be extensible up to some point.

This challenge can be seen in all implementation phases, as outlined here:

- The design work needs to consider that things can be just extended
- The implementation needs to reserve spots for extensions
- The documentation has to mention the extension points
- Those testing need to be aware of them in order to test and solidify them
- Potentially, even a deployment system needs to be aware of them to wire them up correctly
- Teams working on micro frontends have to be disciplined not to violate the core principles that have been decided for the solution
- Additional tools need to be in place to check for violations of the core principles, e.g., whether extension points remain unchanged and are properly used

The end result can be convincing, though. As illustrated in *Figure 4.4*, the split is now functional in nature. Instead of grouping by what is seen on the screen, we actually have corresponding items within the same micro frontend:

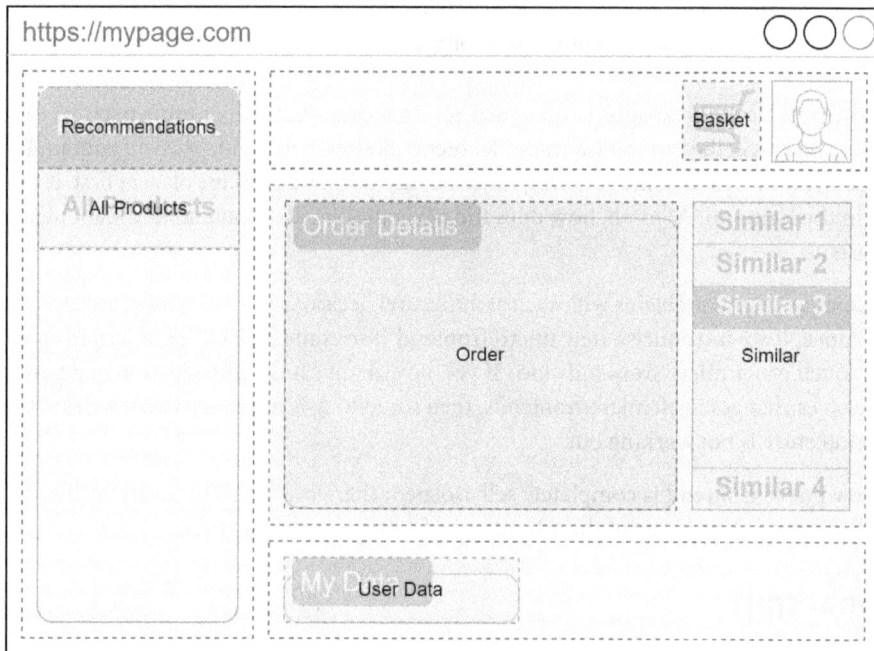

Figure 4.4 – A possible functional split; elements on the screen are extensible and composed from different micro frontends

The great thing about a functional split is that it instantly fulfills the requirement of self-isolation. If we turn off one micro frontend, the others are not directly affected. By comparison, in a technical split, we cannot just turn off the header or footer as it would impact all domains.

Let's take an example of an order micro frontend containing the current order basket component and an order button. Turning this one off will turn off this functionality, but nothing more. The product pages, recommendations, and other parts are all still very much active. Achieving the same kind of implicit flexibility with a technical split is pretty much impossible.

So, grouping by subdomain is not only possible by leveraging a functional split, but actually helpful. This is where DDD becomes a useful tool. Still, while we now know that, for example, an order basket and product recommendations should not be in the same micro frontend, the question is whether the order basket and the order button should be in the same micro frontend.

The question of how small a micro frontend should be is as much up for debate as for microservices. Rather than trying to come up with some kind of metric to measure the size (for example, lines of codes, components, and so on), we should instead try to find a strategy to organize a split. For instance, we should ask the following questions:

- Will the components of the micro frontend all be developed and maintained by the same team?
- Will the components of the micro frontend all be consumed by the same users?
- Will the components of the micro frontend be conditionally hidden or disabled?
- Will the components of the micro frontend be better suited in other collections?

Such questions try to find out if a decision to compose a micro frontend is sound. For instance, if one component should only be used by admins while others are accessible to standard users, it could make sense to factor this one component out into its own micro frontend. Likewise, thinking in terms of user testing—for example, when an A/B test is run to find out whether a certain style of a specific component should be preferred over another—it may make sense to factor out components that should be part of the test.

To make the process of domain decomposition a bit more specific, let's look at an example.

Example decomposition

For this example, we will consider an online shop application. As a requirement, we should deliver the following functionality:

- Login user page
- Login user button
- Logout user button
- Order basket page

- Product details page

- Product overview page

- Special offers page

- Recommendations bar

- Order basket info/link

- Order product button

- Similar products panel

- Product comparison panel

- My account link

- My account page

- Current user info panel

- Important notifications panel

- Customer center page

- Customer feedback modal dialog

- Customer feedback button

Using techniques from DDD, we can arrange these items into different bounded contexts. For instance, we see that we could introduce a user context consisting of the following items:

- Login user page

- Login user button

- Logout user button

- My account link

- My account page

- Current user info panel

Out of this bounded context, we could extract at least two modules, as follows:

- `User login`

- `User account`

The user login module would contain the login page as well as the login button and should only be loaded in the case of an anonymous user. The user account module would contain the rest, including the logout button, and should only be loaded in the case of an authenticated user.

There is, however, a second point of view regarding this bounded context. If we regard the login/logout functionality as something technical rather than domain-specific, we could put all this in a cross-cutting concern category. This way, the bounded context would only consist of the following items:

- My account link
- My account page
- Current user info panel

This bounded context could be implemented in a single module. The remaining functionality would be part of a technical domain that is not implemented in some module, but rather in the application itself.

An advantage of such cross-cutting concerns for micro frontends is that every micro frontend can access them. In the case of treating authentication as a cross-cutting concern, this could mean that every micro frontend can get information regarding the current user, such as a name, email address, or account status.

A disadvantage of cross-cutting concerns is that they represent a shared layer that needs to be maintained and carefully evolved, too. Breaking changes—independently of whether this is directly in the **Application Programming Interface (API)** or the behavior—will have consequences that can only be seen through extensive testing of the entire application.

Coming back to our example, we can identify that there are four more requirements that could be regarded as cross-cutting concerns, as follows:

- Important notifications panel
- Customer center page
- Customer feedback modal dialog
- Customer feedback button

For now, we can divide them into two bounded contexts, as follows:

- User notifications
- Customer interaction

While the notifications panel would be in the user notifications context, the customer interaction could consist of two modules: one containing the customer center page (as well as potential links to it), and another one for everything related to customer feedback.

The remaining items are organized rather quickly. We define bounded contexts for product orders (order basket page, order basket info, order product button), products (product details page, product overview page, similar products page, product comparison panel), and sales (special offers page, recommendations bar). Now, the crucial question comes up: how to decompose these contexts into micro frontends?

Let's pick the bounded context of products to start with. The four items could be placed into the following two modules:

- Product information (product details page, product overview page)

- Product relations (similar products page, product comparison panel)

One possible decomposition is to represent these two modules in two micro frontends. This simple option is definitely a good start; however, it does not mean we are done yet.

As an example, let's pretend that the product comparison page should only be displayed for authenticated users. In this case, it may make sense to actually split the product relations module into two micro frontends: one loaded for every user, and another one only loaded for authenticated users.

Likewise, the product overview page may be a great target for A/B testing. After all, this one will certainly go through a lot of iterations and will impact directly what users are more likely to click on and potentially order. If this is the case, then putting the product overview page into its own micro frontend would be quite beneficial too. This would allow us to load either the product overview micro frontend variant A or variant B per request, supporting A/B testing from the ground up.

As you can see, quite a few decisions here are based mostly on what we anticipate, expect, and require from our application. To technically solidify these expectations, we should introduce clear architectural boundaries first.

Architectural boundaries

When discussing architectural boundaries, we are already entering the code level. Here, we go beyond the domain model and into the following areas:

- What capabilities do the micro frontends have?

- How much freedom do we give the micro frontends regarding their structure and design?

- Which APIs, e.g., of the **Document Object Model (DOM)**, can the micro frontends use?

- Where are the micro frontends decoupled from the application and where are they really tightly bound to the surrounding system?

As usual, the answer is that it depends, but we can still come up with things to consider and keep in mind that they will apply to almost every sound solution.

Let's start with shared capabilities, before determining what is needed for choosing the right level of freedom. After an example of accessing the DOM, we'll conclude with a discussion about the universality of micro frontends.

Shared capabilities

When creating micro frontends, we will feel a tension between two extremes, as outlined here:

- There's (a), putting all shared capabilities and code into a common library or micro frontend
- There's (b), not sharing anything—every micro frontend must duplicate or come up with its own solution

These extremes can be felt on a variety of decisions, but let's focus on the capability decision here.

An advantage of extreme (a) is that every micro frontend can exclusively focus on its problem domain. A disadvantage is that changes to a massive shared code base are likely and these can be destructive, especially when we consider loose coupling between the shared capabilities and the micro frontends.

Knowing that (a) is not the correct approach, we may go for (b); however, with everything just being blindly duplicated, we end up in a similar spot. Once we know some shared capability has a bug, we need to fix it potentially everywhere, leading to massive overhead.

There are two ways out of it: either we go between (a) and (b) by not sharing everything that can be shared, but just some really stable core functionalities, or we go with (a), but still allowing each shared capability to be overridden in the micro frontends. This way, fixes can occur where errors have been seen. Once we know that the same fix needs to be applied everywhere, we could just roll this out everywhere.

So, what could be a shared capability? The following could be defined as such:

- Authentication and authorization
- Permissions and rights
- Feature flags
- Basic user information
- Navigation
- Logging

There is, of course, room for a lot more shared capabilities, but the preceding list should give you an indicator of what could be considered a shared capability.

At this point, it's quite clear that choosing the right level of freedom is relevant, but how do we decide what is the right level? Let's look at that in the next section.

Choosing the right level of freedom

The right level of freedom is mostly determined by the following three factors:

- **Security**—How much do we trust the micro frontends?

- **Performance**—How much should be reused?

- **Team setup**—How are the developers/teams structured?

There is no strict priority here, and other factors may play an important part too. It all depends on the requirements of the project.

In the security area, we need to determine whether micro frontends can be run just like any frontend code or whether they should be particularly sandboxed. Depending on the chosen micro frontend architecture, sandboxing may be more or less problematic. One of the simplest solutions here is to put the micro frontend into an `<iframe>`. For implementers of micro frontends, sandboxing means being restricted in using DOM APIs. It is therefore crucial to explicitly communicate the sandboxing settings as soon as possible.

Security is a lot about trust. If we want to create micro frontends for their compositional advantages but still develop everything in-house within the same team or related teams, then security may not be a major concern. If we want to allow third-party developers to place content on our page, then yes: this is a major concern.

In the performance area, it's all about our desire to have a really fast and swiftly loading application indicated by metrics such as **Time-To-Interactive (TTI)** or **First Contentful Paint (FCP)**. For instance, for an e-commerce website performance would potentially be one of the highest priorities, while for a tool-like portal it would be lower down the ranking. Independent of the priority, to achieve great performance the rendering needs to be as streamlined as possible. This involves choosing the fewest possible frontend frameworks (none, or just a single one) and the fewest number of assets to be required for the initial rendering.

The easiest way to achieve great rendering performance is to use server-side composed micro frontends with a performance budget per micro frontend. Each micro frontend is tested against the preassigned budget and cannot be brought live if it exceeds the set threshold. For other types of micro frontend solutions, the situation is a bit more complicated, but more on this later.

In the team setup area, we need to know what kinds of developers will be writing the micro frontends. Are these likely to be junior developers or really strong developers? For the former, less freedom may be good as it will give them guidance where needed, while for the latter, having more freedom can be beneficial. If we don't know the kind of developer, we should always go for the middle ground, leaning toward the junior developer. Having clear boundaries is also good for senior developers—after all, it's one less thing to think about.

Let's look at one example degree of freedom: accessing the DOM from a micro frontend.

Accessing the DOM

Let's say a micro frontend has some JavaScript included. As we will learn in the next chapter, for client-side composed micro frontends this will always be the case, but even server-side composed micro frontends could come with some JavaScript. In any case, since it's JavaScript, it can do everything, including the following:

- Install a keylogger on the website, sending each keystroke to some server with malicious intentions
- Change all links on the site to potential phishing websites
- Intercept form submissions including a login

Quite horrific, isn't it? To prevent this, we need to limit potential DOM access. As already explained, the simplest solution would be an `<iframe>` whereby we set the sandboxing attributes, which in practice could look like this:

```
<iframe
   sandbox="allow-same-origin
            allow-scripts
            allow-popups
            allow-forms"
   src="/some-microfrontend.html"></iframe>
```

Once we use the `sandbox` attribute, everything is disallowed, meaning that every feature needs to be turned on explicitly. In this case, the micro frontend would be allowed to communicate with the same origin as the parent frame, load JavaScript, use popups, and allow the submission of forms.

There are two other things that we can consider helpful to improve security. If we think something is entering on the user's side (such as in a **man-in-the-middle** (**MITM**) attack), we can use **Subresource Integrity** (**SRI**) to prevent this.

> **Important note**
> SRI is a security mechanism that blocks resources from being integrated after manipulation. It works by specifying a cryptographic secure hash on the `integrity` attribute of the declaring element; for example, a `<script>` tag. The browser then compares the hash of the downloaded content with the specified value. More information can be found at `https://developer.mozilla.org/en-US/docs/Web/Security/Subresource_Integrity`.

Another common way to limit access is to use **CSP**, which stands for **Content Security Policy**. Here, a **HyperText Transfer Protocol** (**HTTP**) header defines what is actually allowed and what is not allowed. Since CSP is all about content, we cannot really limit the DOM API; however, we can limit the origin of the different kinds of resources (for example, images, style sheets, documents, and so on). We can also actively prevent inline scripts, which includes the use of unsafe APIs such as `eval`.

Finally, another thing we could do is to allow scripting, but only within a web worker. To ensure this, we will manipulate the **HTML** before sending it out and replace the script with a small inline script, which creates a web worker and does some of the allowed interaction. This method assumes that we have access to the HTML and does not work if the micro frontend is served from elsewhere.

The code for the inline script could be as simple as the following example:

```
if ('Worker' in window) {
  var worker = new Worker('/previous_script_url_here.js');
  worker.onmessage = function (e) {
    try {
      var msg = JSON.parse(e.data);
      switch (msg.type) {
        case 'change_text';
          var s = document.querySelector(msg.selector);
          s.textContent = msg.text;
          break;
        // more APIs here.
        default:
          console.warn('Unrecognized message type.', msg);
          break;
      }
    } catch (ex) {
      console.warn('Unrecognized message format.', ex);
    }
  };
}
```

In the example, we used **ECMAScript 5 (ES5)** with an API check to prevent failure on older browsers. If we use this technique, we also need to communicate the replacement and API properly.

The level of freedom also influences how easily micro frontends can be transported from one system to another—in other words, how reusable they are.

Universality of micro frontends

The reusability of micro frontends is something that I would put under the *universality* category. If a micro frontend is really usable in a broad way—that is, by another system that comes with similar boundary conditions but otherwise has nothing in common—we have a fortunate situation. In this scenario, we create a higher-order building block that can be used in different contexts. Consider multiple such building blocks, and we can assemble new web applications in practically no time.

In order to achieve this universality, however, we need to find the right balance regarding the requirements for running our micro frontend—for instance, shared capabilities that need to come from the surrounding system should be minimized. Any shared capability places a requirement on the system where the universal micro frontends should be run.

Likewise, from a security point of view, we should regard the surrounding system as very strict. After all, since we refer to universal micro frontends solutions, we don't know the potential target systems, making them as strict as possible. We can soften all these requirements once we drop the universality claim and reduce ourselves to a few known target systems.

Besides some elementary requirements, universal micro frontends contain all functionalities. This makes universal micro frontends a great candidate to expose frontend-driven **Software-as-a-Service (SaaS)** offerings. Examples in this space include chatbots, quick user feedback provision, and cookie consent dialogs. All of them are implemented as universal micro frontends, and most of them are implemented via loader scripts, which will create an `<iframe>` where the frontend is hosted.

Summary

In this chapter, you learned that conceptual alignment and organization are as crucial—if not more crucial—as the actual implementation of micro frontends. After all, the domain decomposition gives us the boundaries that are needed to organize teams, define APIs, and assign functionality to the different modules.

You learned some basic vocabulary from DDD and how a context map can help you to organize the different bounded contexts that have been derived from the identified subdomains. You've seen that DDD tries to put everything into smaller chunks using a tech-agnostic language, such that the implementation can be chosen independently.

You also learned that a clear SoC and strict architectural boundaries lead to more fine-grained modules with clear team responsibilities.

In the next chapter, we will get an overview of the available types of micro frontends, which types exist, and when they should be used.

Part 2:
Dry Honey – Implementing Micro Frontend Architectures

In this part, you will gain in-depth knowledge on the available architecture patterns, their implementation, variations, application, maintenance, advantages, disadvantages, and common challenges.

This part covers the following chapters:

- *Chapter 5, Types of Micro Frontend Architectures*
- *Chapter 6, The Web Approach*
- *Chapter 7, Server-Side Composition*
- *Chapter 8, Edge-Side Composition*
- *Chapter 9, Client-Side Composition*
- *Chapter 10, SPA Composition*
- *Chapter 11, Siteless UIs*

5

Types of Micro Frontend Architectures

I hope you are now convinced that micro frontends may bring something good to your projects, but some work needs to be done before we can actually start implementing them. As you've learned in the previous chapter, decomposing the domain properly is the most important aspect.

Now that we've done our homework regarding the basics, it's time to look at the implementation options. Sometimes, it may be quite obvious how to implement our application; however, most of the time, it makes sense to check against some existing guides and best practices first.

There are many different types of micro frontend architectures. Compared to microservices, the situation feels much more fragmented for micro frontends. One reason for this fragmentation is that the frontend gives us more options to play with. For instance, we could render everything server side but also do this on the client side. We could also go for a mixed approach.

In this chapter, we'll first try to understand the current landscape of micro frontend implementations by exploring several key types. This includes static, dynamic, horizontal, and vertical splits, each with its own applications and challenges. Then, we'll cover each of the three key implementation properties specifically.

This will give us a kind of decision tree that we can use. We'll have three core decisions and end up with seven unique solutions. Each solution is discussed with its pros and cons, areas of use, and some high-level implementation details.

Here are the topics we will be covering in this chapter:

- The micro frontends landscape
- Static versus dynamic micro frontends
- Horizontal versus vertical split
- Backend- versus frontend-driven micro frontends

Let's start by acquiring some knowledge of the micro frontend landscape.

Technical requirements

To follow the code samples in this book, you need knowledge of JavaScript and how to use the command line. You should have Node.js (version 20 or higher) installed using the instructions at `https://nodejs.org`.

The code used in this chapter can be found in the following repo:

`https://github.com/PacktPublishing/The-Art-of-Micro-Frontends-Second-Edition/tree/main/Chapter05`

The CiA video for this chapter can be found at `https://packt.link/DeGj5`

The micro frontend landscape

Either you know this already or you'll realize it sooner or later reading this book: micro frontends are not a new idea. Actually, they are a very old idea, dating back to before microservices and even **Service-Oriented Architectures (SOAs)** were formed. It's just that the possibilities changed, as well as the expectations. Giving an old idea a new name is an easy trick to raise interest, and it worked.

Going beyond the web, we see plugin architectures as a very common pattern in many **user interface (UI)** technologies. Actually, some very popular applications such as Microsoft's Office applications used them already very early on. While there are some differences to micro frontends, in the end, they all have similar goals and challenges have been recorded. As a result, plugin architectures have been popular in web frameworks early on too.

Today, plugin architectures are still super-popular and provide an essential point to almost any larger-scale application, framework, or tool. Where would **Visual Studio Code (VS Code)** be without extensions? Can you imagine webpack without a loader or plugin? What if Babel did not allow presets or plugins?

While many plugin architecture implementations are runtime mechanisms today, some of them require a restart of the application, or even a recompilation. The interest in micro frontends, however, did not grow due to it using a single software design pattern, but it was rather influenced by its openness towards the used implementation.

As far as potential implementations are concerned, there are at least three core decisions that influence the type of micro frontend to choose. We need to know the type of dynamic with respect to the usage of micro frontends, our team setup, as well as the place where we want to make the composition.

The result is a **three-dimensional (3D)** phase space for categorizing micro frontends. The phase space contains dimensions tied to the composition time (from static, i.e., at build-time, to dynamic, i.e., at runtime), the composition host (either on the client or on the server), and the composition type (horizontal means one micro frontend across multiple pages, while vertical means one micro frontend per page).

An illustration of the 3D phase space is shown in *Figure 5.1*.

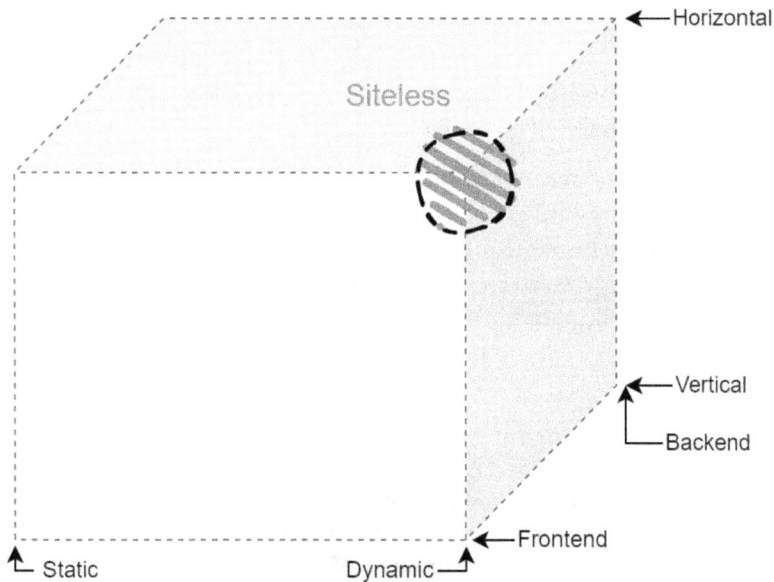

Figure 5.1 – The phase space for categorizing the types of micro frontends; patterns
such as Siteless appear depending on the values of their key properties

The phase space can be filled with the patterns we'll introduce in the next chapters. The location of these patterns is no exact science. In a specific implementation, we could always try to bring the value from one extreme to a more generic middle value.

For instance, while this example shows **Siteless** as a dynamic, horizontal architecture that will be composed on the frontend, it could also be either fully rendered on the backend or be enhanced with a bit of **server-side rendering** (**SSR**). Either way, it would be shifted to the right. Likewise, we could compose teams a bit differently to lower the marker a bit.

Let's start our discussion of these extremes with our first axis—static versus dynamic micro frontends.

Static versus dynamic micro frontends

One of the simplest approaches for implementing micro frontends is to decompose a development into several packages that are then brought together at build time. This is, however, a fully static usage of micro frontends.

The main advantage of a static approach is that all information is known at build time and can therefore be used by the tooling. This leads to possible optimizations, deeper integrations, and enhanced checks. Faster and more reliable applications can be realized using this approach.

The main disadvantage of a static approach is that a significant change in any micro frontend—such as an addition or a removal—requires a change to the main application. Furthermore, any change will trigger a rebuild of the full application.

The primary use cases of static micro frontend solutions are slowly changing websites or smaller web applications. One example framework here is Bit (`https://bit.dev/`).

In the easiest case, a static micro frontend solution just includes various packages with a single entry point. Let's consider a straightforward example using Node.js with Express to illustrate this. We'll set up and use a **mono repository** (**monorepo**) to transport one server solution coming with multiple micro frontend packages. The idea would also work without a monorepo. In this case, we'd use packages distributed on a private or public **npm** registry.

> **Important note**
>
> The public npm registry is located at `https://registry.npmjs.org` and allows packages to be published for free. The downside is that all packages are then publicly available. Some **continuous integration** (**CI**) providers such as Azure DevOps give you a free private npm registry, too. Alternatively, open source solutions such as Verdaccio can be used to host a simple npm registry server in any environment. More information on this can be found at `https://verdaccio.org/`.

To create this solution, we need to run the following commands:

```
# Create the Node project
npm init -y


# Make it a Lerna monorepo
npx lerna init --packages="packages/*"


# Add the application itself
npx lerna create @aom/app --yes
npm init -w packages/app -y


# Add some (e.g., 2) micro frontends
npx lerna create @aom/mife-1 --yes
npm init -w packages/mife-1 -y
npx lerna create @aom/mife-2 --yes
npm init -w packages/mife-2 -y


# Register the dependencies
npm i @aom/mife-1 -w @aom/app
npm i @aom/mife-2 -w @aom/app
npm i express pug
```

These dependencies allow us to have a web application using the express library for Node.js with pug as a dependency used as a view engine within Express.

Now, the main application may or may not register some routes and application fundamentals. Keeping it to the point, we end up with the following code to initialize Express and set everything up:

```
const express = require('express');
const app = express();
const port = process.env.PORT || 1234;

app.set('view engine', 'pug');

// just an index page "/"
app.get('/', (_, res) => {
  res.render('index', { title: 'Sample', message: 'Index' });
});

// set up the micro frontends
require('@aom/mife-1')(app);
require('@aom/mife-2')(app);

app.listen(port, () => {
  console.log(`Running at ${port}.`);
});
```

The other pages are all integrated from the micro frontends. Their integration point is a simple setup function. For micro frontend mife-1, it looks like this:

```
const { join } = require('path');
const express = require('express');

module.exports = setupMicrofrontend1;

function setupMicrofrontend1(app) {
  const publicDir = join(__dirname, '..', 'public');
  const middleware = express.static(publicDir);
  app.use('/mf1', middleware);

  app.get('/mf1', (_, res) => {
    const page = require.resolve('../views/index.pug');
    const props = { title: 'Sample', message: 'MF1' };
    res.render(page, props);
  });
}
```

This gives us everything we need. We have an integration point but are pretty much self-contained in our micro frontend. However, all we can do here is create handlers for full **Uniform Resource Locators (URLs)**. We'll see in the next chapter that this can indeed be a suitable basis for a micro frontend implementation.

One problem with the preceding approach is that we need to be careful with the paths—for example, for the views. In a monolith, we could only throw around relative URLs and make them work against a `views` directory. Since we are now in our own package running within some other application, we cannot do that.

Unfortunately, without much work, there is also no way of having a path resolution that is context-sensitive. For now, the approach is to use absolute paths instead.

The beauty of a static approach is that it is just pretty much usable after creating a new micro frontend by using some boilerplate code. There is no lengthy infrastructure configuration or integration fine-tuning, yet in contrast to a dynamic approach, any change to a micro frontend will require a rebuild of the full application.

> **Important note**
> The technique of starting a new project with a boilerplate or template code base is quite popular. Usually, this technique is known as scaffolding, which takes into consideration the transfer of custom properties to a known working solution, following best practices to emerge as a new project.

On the other side, a dynamic approach is much more difficult to implement. There are three challenges to be tackled, outlined as follows:

- How to publish micro frontends to a source
- How to update a source
- How to connect an application to a source

The main advantage of a dynamic approach is that micro frontends can be selected on a per-request level, which gives developers a lot of freedom. Also, updates of micro frontends can just happen continuously without interrupting the main application.

The main disadvantage is that the complexity and loose coupling will lead to more fragile applications. Additional tools and error boundaries are here to help but will also add more complexity on the infrastructure level.

The primary use cases of dynamic micro frontend solutions are personalized websites or larger web applications. One example framework here is **Module Federation**, which started as a plugin for the webpack bundler.

While the decision between dynamic and static micro frontends is mostly answered by a project's requirements, other decisions are made by looking at the anticipated team structure. An example of this is the decision between a horizontal split and a vertical split.

Horizontal versus vertical split

Micro frontends can be created on a per-view basis, as well as in a compositional way where multiple teams contribute to the same view. Using micro frontends per view, we refer to a vertical split, while the composition of multiple micro frontends per view is called a horizontal split.

The following screenshot illustrates a typical vertical split architecture:

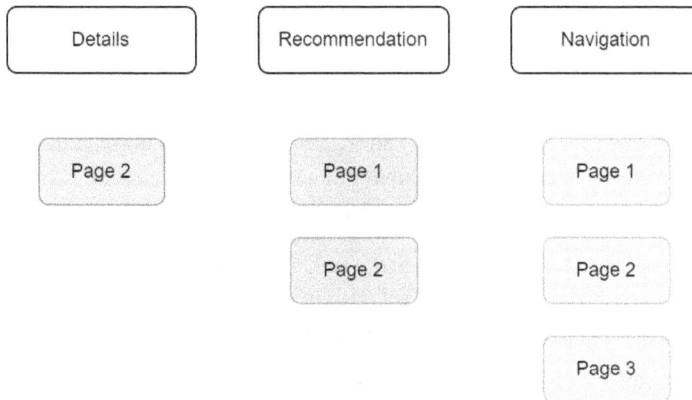

Figure 5.2 – In a vertical split, multiple teams deliver multiple pages, each composed of multiple features

The main advantage of a vertical split is that it is very easy to reason with. Every micro frontend is not only developed in isolation, but is also most often found as an isolated web application in one way or another.

The main disadvantage of a vertical split is that it does not scale well. While most websites use parts from multiple subdomains on some pages, the vertical split does not advocate combining micro frontends to allow this.

The primary use cases of vertical split micro frontend solutions are content-heavy websites or single-use-case pages. One example framework here is **Podium**.

In contrast, a horizontal split uses true cross-functional teams to develop micro frontends that should only require knowledge of a single subdomain. Here, the complexity lies mostly in providing a system that can be debugged and extended well.

The following screenshot illustrates a typical horizontal split:

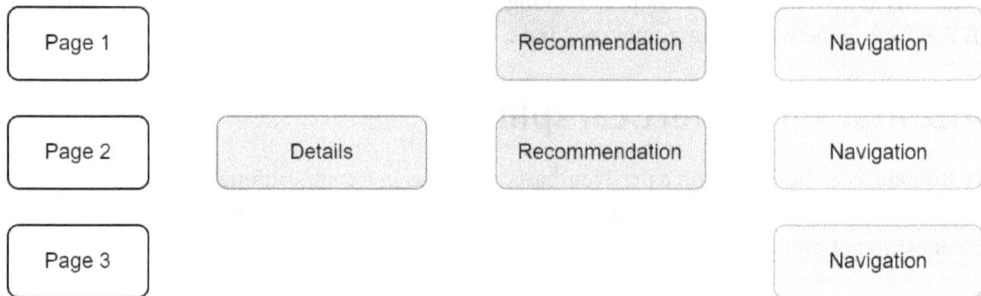

Figure 5.3 – In a horizontal split, multiple teams deliver multiple
features, each scattered across multiple pages

The main advantage of this approach is that we can truly split a problem domain into smaller parts as we wish, allowing us to focus on a single subdomain in a single micro frontend. Consequently, some pages are composed of multiple micro frontends.

The main disadvantage of the horizontal split is that a lack of focus on providing ready-made pages will make things complicated for developers. This will also impact their ability to debug the application, as many screens require debugging of multiple micro frontends.

The primary use cases of horizontal split micro frontend solutions are large web applications, as well as web portals. One example framework here is **Piral**.

The horizontal and vertical splits present a valid distinction that may play an important role when selecting the type of micro frontend architecture. Even more important is the decision between backend and frontend as the area of composition. There are multiple reasons for this. First, their implementations are significantly different. Second, they have different limitations and challenges – and consequently different application targets.

Backend- versus frontend-driven micro frontends

Very often, a micro frontend approach discussion starts with the almost classic "**server-side rendering (SSR)** versus **client-side rendering (CSR)**" topic. While some of the arguments would apply to a monolith too, other arguments that only apply to micro frontends can be found as well.

> **Important note**
>
> The discussion of SSR versus CSR is a relatively new one. Only since JavaScript frontend frameworks have become powerful enough that they not only *can* be used for CSR but actually *should* be used for it have people actively leveraged this option, and, indeed, in many cases, the option is not only quite simple but also has the least infrastructure complexity and great scalability. Nevertheless, for really fast websites, either pre-rendering or a mix of SSR and CSR will hit the sweet spot.

Backend micro frontends, often called server-side micro frontends, were among the first types of micro frontend implementations. One reason is that SSR is the original way of enabling dynamic websites. Another reason is that the necessary technology has long since been ready for micro frontends. With **Server-Side Includes (SSI)** and its successor **Edge-Side Includes (ESI)**, these two possibilities have existed for a long time.

The main advantage of a backend approach is that the delivery of the micro frontends cannot be more frictionless for the end user. In most cases, the perceived performance is in the same ballpark as for monoliths. Furthermore, for non-JavaScript users and robots, the delivery already contains the core information, which is great for SEO.

The main disadvantage of a backend approach is that we need a quite complicated infrastructure to sustain scalability and reliability.

The primary use cases of backend micro frontend solutions are e-commerce websites and content portals. One example framework here is **Mosaic 9**, which, unfortunately, is no longer actively maintained.

Let's come back to our previous sample code for a build-time integrated micro frontend solution. We can modify this to a server solution with dynamic routing of views. In this case, the application is taking the role of a kind of micro frontend view gateway, while each micro frontend is no longer a package, but rather its own application running on a dedicated port.

In other words, we'll transform the following:

- `app` to be an application gateway running on port `1234`
- `mife-1` to be its own application running on port `2001`
- `mife-2` to be its own application running on port `2002`

For illustration purposes, the lookup in the application gateway will remain static right now. We'll make it more dynamic and powerful in the upcoming chapters.

In the case of two micro frontends, the static lookup is quite straightforward, as can be seen in the following code snippet:

```
const lookup = {
  ,/mf1': `http://localhost:${process.env.MF1_PORT ||
    2001}`,
  ,/mf2': `http://localhost:${process.env.MF2_PORT ||
    2002}`,
};
```

By using a proxy solution such as `http-proxy`, we're able to just forward the requests to the destination. We'll do that dynamically, based on the actual contents of the `lookup` variable, as follows:

```
const proxy = httpProxy.createProxyServer();

app.use((req, res) => {
  const [prefix] = Object.keys(lookup).filter((m) =>
    req.path.startsWith(m));

  // nothing found, let's just return an error page
  if (!prefix) {
    return res.status(404).send('Nothing found.');
  }

  const target = lookup[prefix];

  // let's proxy the request - it should be fully handled
  // on the respective micro frontend's server
  return proxy.web(req, res, { target }, e => {
    console.error(e);
    res.status(500).send('Something went wrong...');
  });
});
```

Make sure to handle errors in the proxy connection gracefully. The last thing we want in a micro frontend solution is for one of the micro frontends to bring down the whole application.

For completeness, let's have a look at one of the transformed micro frontends. It is now fully self-contained and could be debugged individually, too. You can view this in the following code snippet:

```
const path = require('path');
const express = require('express');
const app = express();
const port = process.env.MF1_PORT || 2001;
```

```
app.set('view engine', 'pug');

app.use('/mf1', express.static(path.join(__dirname, '..',
  'public')));

app.get('/mf1', (_, res) => {
  res.render('index', { title: 'Sample', message: 'MF1' });
});

app.listen(port, () => {
  console.log(`Running at ${port}.`);
});
```

Right now, the two things that are still quite annoying are the implicit knowledge sharing (that is, the chosen route prefix needs to be known in the micro frontend as well as in the application gateway) and the build-time coupling due to the monorepo. The latter, however, is only an artifact from the first prototype and is actually quite easy to remove.

At this point, the server-side-composed micro frontend solution seems to be quite appealing. After all, we get things such as an isolated debugging experience out of the box. Furthermore, we may be able to transport some of our microservice knowledge into our micro frontend infrastructure.

On the other hand, client-side micro frontends seem to be more in demand these days, one reason being that they present a more direct approach. Seeing micro frontends as a real frontend game changer, it makes sense to apply this architecture without any backend requirements at all. Quite often, unfortunately, this overlooks many of the great optimizations and enhancements that are offered exclusively by mixing in some backend capabilities. Taking the best of both worlds is where we'll find the ideal solution for most problem descriptions.

The main advantage of a frontend approach is that it allows the most flexibility of all approaches. After all, any UI framework can be used here, no matter whether it renders exclusively on the server side or exclusively on the client side.

The main disadvantage of a frontend approach is that the assembly and composition will always take quite some time. Being quite wasteful on resources is not ideal when trying to avoid a scalability bottleneck.

The primary use cases of frontend micro frontend solutions are tool-like experiences and web applications. Two example libraries here are **single-spa** and **Picard**.

With these three key decisions in mind, it is now time to leave the theoretical area behind and start focusing on actually implementing micro frontends.

Summary

In this chapter, you've learned that there are multiple types of micro frontends. Depending on the project requirements, the selection of one of these needs to be made quite carefully.

You've learned that there are certain key properties that define the potential solution space, especially the three areas of build time versus runtime, per-view versus in-view, and on-server versus on-client matter.

You've seen that each decision leads toward a potential implementation approach. While the exact implementation approach is yet to be defined, the primary area of use for each of these types should be quite clear right now.

In the next chapter, we'll start with our first micro frontend implementation. We'll use a classic web approach to integrate multiple web servers into a single web application.

6

The Web Approach

In the previous chapter, you saw that we can construct many different types of micro frontends. All we need to know are the boundary conditions of our system. These core technical requirements play a crucial role when we go in favor of or against a potential technical solution.

Now it's time to get our hands dirty. From here on, we'll live in code. We will now implement the most widely spread architecture patterns for implementing large micro frontend systems. We'll start with the simplest pattern and move on until we reach the Mount Olympus of micro frontend architectures.

In this chapter, we'll introduce the web approach as the most basic pattern to implement micro frontends. This will be a refinement and drastic enhancement of the previous sample code, illustrated in the backend section of the previous chapter.

We'll start with some of the basics of this architecture pattern. Here, we will also introduce a sample implementation – still working in a single repository, but actually simulating multiple repositories to see how this would play out.

Afterward, we'll discuss the advantages and disadvantages of this pattern in full detail. Finally, we will go into two specific enhancements that make this pattern applicable: how to dynamically generate the links for the individual micro frontends and how to leverage iframes fully.

In short, we'll touch on the following topics in this chapter:

- Basics of the web approach
- Advantages and disadvantages of the web approach
- Using links for navigation
- Using fragments with iframes

Without further ado, let's jump right into the topics.

Technical requirements

To follow the code samples in this book, you need knowledge of JavaScript and how to use the command line. You should have Node.js (version 20 or higher) installed, using the instructions at `https://nodejs.org`.

The code used in this chapter can be found in the following repo:

`https://github.com/PacktPublishing/The-Art-of-Micro-Frontends-Second-Edition/tree/main/Chapter06`

The CiA video for this chapter can be found at `https://packt.link/0gLGm`

Basics of the web approach

The web approach of implementing micro frontends works by referring to individual micro frontends via their URL. The basic principle is shown in the following diagram:

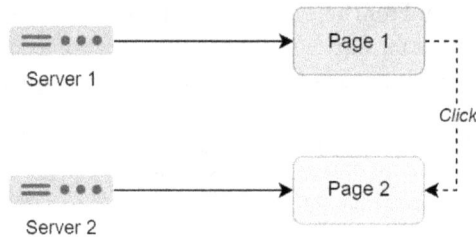

Figure 6.1 – Idea behind the web approach – different servers providing
different parts of the application combined with hyperlinks

In this pattern, teams go for whole pages and are governed by a central design and business domain.

In practical applications, following the basic principle, this could mean that the origins, i.e., the protocol and domain of the website, change during navigation, i.e., when a user follows a link. For instance, a user could start at `www.example.com` but then navigate to `mf1.example.com` without leaving the actual website.

More often, however, it means that the real web servers are hidden behind a kind of gateway, as implemented in our previous sample. In this case, the origin would remain the same, but the paths may change – for example, navigating from `/` to `/mf1`.

Let's look into the architecture of the web approach before doing a sample implementation. Finally, we'll also discuss some potential enhancements that would improve the sample implementation.

The architecture

At its core, the web approach uses an architecture that is pretty much the same as microservices, where the different modules are deployed as individual services – each with its own URL. However, instead of serving pure data in forms such as JSON, a presentation using HTML is already rendered.

The web approach does not dictate how exactly this is implemented. For instance, we could just leverage an existing microservice backend and transform the services to be sensitive to the `accept` header.

Seeing a value such as `text/html` as top priority, we'd send back the presentation version:

```
Accept: text/html, application/xhtml+xml,
    application/xml;q=0.9, image/webp, */*;q=0.8
```

Alternatively, we could set up dedicated web servers used for the individual micro frontends, but using the existing API servers to retrieve the data. This may be the weapon of choice for purists, but it will also come with the highest infrastructure demands.

Another thing to consider is an aggregation layer such as an API gateway. We already used this enhancement in the former sample and we will advocate it in this chapter again. The reason is simple: it avoids a lot of issues by communicating to the browser that all these pages run within one larger application.

From the perspective of a browser, the application boundaries are defined by a lot of things, but the most important one is the website's origin, which is the part of the URL before the path information.

In our sample implementation, we'll also go for an aggregation layer.

Sample implementation

Previously, we always went for a monorepo; however, for this sample, we will actually simulate a setup where each micro frontend lives in its own repository.

The reason is simple: in the real world, at least some micro frontends will be developed in their own repositories. The freedom of choice and the possibility of external development are two of the important factors when considering micro frontends. It's much easier to have a setup that can work with distributed repositories than a setup that can work with a monorepo.

We start by creating a dedicated space for each micro frontend. Usually, these would be separate repositories, but for this sample, it's enough to create separate directories.

In the web approach, every micro frontend needs to be a full web server. It can be a dynamic page using something such as PHP or Node.js, or just some static HTML markup. What we can do for demonstration purposes is to go for Node.js projects, which just use the `http-server` package for serving static HTML.

A micro frontend can therefore be fully scaffolded using the following:

```
# Initialize a new Node.js project
npm init -y
# Add the http-server package to the dependencies
npm i http-server --save
```

The `scripts` section of `package.json` needs to be extended with a `start` script:

```
"start": "http-server ./views --port 2001"
```

This command serves the `views` directory at `http://localhost:2001`. We can run it via `npm start`.

The actual HTML is rather boring, as it's not any different from the HTML we would write for a non-micro frontend page. The only difference is that we use links that refer to other micro frontends – for example, our first micro frontend may contain a link such as the following:

```
<a href="/mf2">Go to MF2</a>
```

As we will discuss later, these links are rather fragile and should potentially be made more robust.

For our sample, we will also add an aggregation layer. This is just a micro frontend proxy. There are many possible technology choices here. In this case, we will go for a Node.js server using `Express` with the `http-proxy-middleware` package.

One way to implement the gateway is to forward / to some micro frontend while proxying every request based on the path prefix. This sets the requirement on every micro frontend to use a certain prefix – and stick to it. If no known prefix is identified, we'll respond with the HTTP status `404`.

The code block will look something like this:

```
const express = require('express');
const { createProxyMiddleware } = require('http-proxy-
  middleware');

const app = express();
const port = process.env.PORT || 1234;

const targets = {
  ,/mf1': ,http://localhost:2001/mf1',
  // ... more
};

app.get('/', (_, res) =>
  res.redirect(Object.keys(targets)[0]));
```

```
Object.keys(targets).forEach((prefix) => {
  app.use(
    prefix,
    createProxyMiddleware({
      target: targets[prefix],
      changeOrigin: true,
    })
  );
});

app.get('*', (_, res) => res.status(404).send('Page not
  found.'));

app.listen(port, () => {
  console.log(`Gateway running at ${port}.`);
});
```

The directory structure of the sample implementation is as follows:

```
mf-1/              # repository of micro frontend 1
├ views/          # views provided by the micro frontend
│  ├ mf1/         # folder present to expose right namespace
│  │  ├ index.html     # main view of micro frontend 1
│  │  ├ fragment/      # expose additional fragment
│  │  │  ├ index.html  # fragment view from micro frontend 1
│  │  │  ├ packages.png # resource used in the fragment
├ package.json # project details
mf-2/              # repository of micro frontend 2
├ views/          # views provided by the micro frontend
│  ├ mf2/         # folder present to expose right namespace
│  │  ├ index.html     # main view of micro frontend 2
├ package.json # project details
mf-gw/             # repository of the aggregation layer
├ package.json # project details
├ lib/            # runtime files
│  ├ index.js # server startup file
```

While it could be argued that some crucial features are missing to achieve production readiness, there are other potential enhancements that should be considered too. Let's look at them in the next section.

Potential enhancements

We've already introduced the gateway as a potential enhancement. If we introduce such a central point, we should make sure not to strongly couple all the web servers to it. Therefore, some loose coupling via a dedicated registry service or another discovery mechanism is strongly recommended.

Out of the box, the web approach is already quite robust. It also comes quite loosely coupled, which is a good thing. Nevertheless, in order to enforce certain patterns, we may want to create a boilerplate including common things such as logging, error handling, or authentication. If we have this boilerplate, individual teams can use it as a starting point. This would boost productivity immensely.

Another thing to consider is a service to enforce certain UX patterns or other key properties that we want to rely on. For instance, determining a performance budget per micro frontend is a good start, but without a tool to enforce it, is quite meaningless.

What other advantages and disadvantages exist when using this pattern? Well, let's take a look at them in the next section.

Advantages and disadvantages of the web approach

The major advantages of the web approach are the similarity and pureness in relation to microservices, as well as its simplicity. Like plain microservices, no additional effort, libraries, or infrastructure is needed – that is, no fancy JavaScript techniques, frontend frameworks, or requirements on the actual web server technology.

The major issues also inheritably come with these advantages. For instance, since the frontend has a much higher need for consistency using common UX patterns, it's quite difficult to stay close to microservices. These are, by definition, not necessarily consistent. Likewise, the simple approach will hit limitations quite fast.

Looking at the limitations, we see that there are only two ways of referring to or using different micro frontends: links (for full-page transitions) and iframes (for UI fragments). We will discuss both in full detail later in this chapter.

So, when should the web approach be used? The web approach makes sense when consistency is not necessarily required. It also makes sense when fragments should either be included from third-party sources or when fragments should be provided to third-party targets. Consequently, most websites using third-party cookie consent solutions, chatbots, or similar services will leverage micro frontends already.

This pattern is also a classic in combination with other approaches. For instance, even larger websites such as Amazon use it to put individual applications (for example, Amazon Music, Amazon Video, Amazon Photos, Amazon Shopping) under one umbrella. The different applications don't need to be fully consistent. Instead, their job is to appear as one large product from a single company.

Since everything relies on links so much, we should have a closer look at their use in this pattern.

Using links for navigation

The central mechanism for the web approach is the usage of hyperlinks. Hyperlinks use URLs to refer to resources such as full pages or assets such as style sheets or JavaScript files.

Hyperlinks may seem simple at first, but they are the magic ingredient that has made the World Wide Web (often just referred to as "the web") a success. Due to their loosely coupled nature, they have, however, one big downside: there is no direct way to keep a hyperlink working.

As an example, let's say we are on a page from *micro frontend 1* and want to link to some page from *micro frontend 2*:

```
<a href="/mf2/some-page">More details</a>
```

The problem with this code is that we need to have and use some information from another micro frontend – the URL of the page. Now, the information may be correct when we introduce the link, but we cannot possibly guarantee that for the full lifetime of the link. As such, this is a rather fragile construct.

The origin of the problem is that we use information from another source that can change at any point in time. The question now is: can we change this somehow?

Obviously, the direct answer is no. The link is always determined by its owner. However, what we can do is introduce either a central or a local linking directory. The idea behind such a linking directory is that target links may change, but this change can be observed and corrected on the fly. Consequently, this should act as a kind of reliability layer. Let's take a look at central and local linking directories in more detail next.

Central linking directory

In a central linking directory, we'll store, find, and validate hyper references in a central service. This could be a simple CRUD service or an integrated part of the gateway server.

The idea behind the linking directory is that each micro frontend needs to report all its URLs. This means that new URLs, removed URLs, and changed URLs will be known, too. The latter is especially useful. If each micro frontend can introduce aliases (or dependencies) on existing URLs, we can update the aliases.

Let's say we map a URL, `/common/mf2-some-page`, to `/mf2/some-page`. If `/mf2/some-page` changes to `/mf2/another-page`, the alias will remain the same, but the target will be updated. This way, we have stable URLs without preventing refactoring or blocking changes.

One problem of a central approach is that the individual, distributed micro frontends require a service out of their scope. Such dependencies may become problematic for robustness and debugging reasons. If this is indeed the case, a local linking directory could be a suitable solution.

Local linking directory

Instead of using a central service, we can also integrate a local directory into the web server of our micro frontend. The big advantage of this is that it remains local. The disadvantage is that the infrastructure is more complicated. Here, we require a synchronization or explicit validation mechanism to ensure robust URLs.

In *micro frontend 1*, we could introduce alias URLs such as /mf1/mf2-some-page, which redirect to the actual URL (for example, /mf2/some-page). Now, if this relation is hardcoded, we hardly get any benefits (except that we only need to change a single location instead of identifying where the HTML code refers to the given URL). However, there is a trick.

If each micro frontend comes with a local linking directory, we can expose this directory as an API. Now the directory can actually have two parts: one part to expose all links using unique IDs and the other part to actually show the external links with their current values.

A request to /mf1/links could yield the following response:

```
{
  "internal": [
    {
      "id": "fb2048fb-470c-48f3-85fe-01645adfcd0f",
      "url": "/mf1/first"
    }
  ],
  "external": [
    {
      "id": "d08ee9a5-5f04-41eb-b1c7-2f1a917c1f57",
      "alias": "/mf1/mf2-some-page",
      "url": "/mf2/some-page"
    }
  ]
}
```

This allows periodic checks on the respective service using the given ID. The crucial part here is that IDs should never be changed. They may be removed to indicate a removed URL, but they may never be changed. The url value may change, but this is why a fixed ID is used in the first place.

Having a linking directory not only helps with hyperlinks but also with references to assets. This is quite handy for iframes, too. So, let's move on to iframes now.

Using fragments with iframes

In the web approach, fragments are included via `<iframe>` tags. The only requirement for such a tag is a URL that leads to the fragment. Since the URL of a fragment has to be treated like any URL – for example, hyperlinks, we also need to follow the same principles as outlined in the previous section. Likewise, to avoid using URLs that are fully owned by other teams, we should go via the global linking directory, too.

The following is an example of a fragile reference (from *micro frontend 2*):

```
<iframe src="/mf1/fragment"></iframe>
```

A better way would be to change that to the following:

```
<iframe src="/mf2/mf1-fragment"></iframe>
```

Here, the server would usually just respond with HTTP status `304` redirecting to `/mf1/fragment` using a local linking directory, as introduced beforehand.

There are still some challenges with inline frames. One that has been touched on already is security, while others are accessibility and layout. Let's outline these three.

Security

Besides making the source links more flexible and robust, we also need to think about the right security settings. As discussed earlier, in *Chapter 4*, we can use the `sandbox` attribute to properly protect our top frame from the child frame's content.

Accessibility

While security is one of the advantages of an `<iframe>`, the accessibility and **search engine optimization** (**SEO**) features are quite insufficient – especially if these are placed on customer-facing sites. Screen readers often have quite a problem analyzing a page's content. Since the markup does not provide any assistive semantics, other tools may be less useful, too.

Layout

Since inline frames are just embedded in another document, they cannot influence their own styling or position. Sometimes, this is the desired behavior, but most often this comes at a cost. An inline frame cannot communicate its own dimensions and properly reserve space for them. As such, the parent frame needs to do that.

This is a potential blocker. On the one hand, the parent frame should have the least knowledge about the embedded content, and on the other hand, it needs to know its dimensions upfront to properly include it.

If we fail to accommodate the appropriate space, we end up with unwanted scroll bars or too much white space. Keeping the CSS up to date is, however, not an easy task.

One way to circumvent this is to use a `<script>` instead of an `<iframe>`. Here, we would still ultimately include an inline frame; however, this would not be done in HTML directly but via JavaScript manipulation of the DOM.

The advantage of this approach is that the script would live in the parent document and thus could do the styling. However, the disadvantage is that now the `<iframe>` would only be shown when JavaScript is active. Surely, we could include it via a `<noscript>` tag too, but under normal circumstances, the frame would load more slowly than in a direct approach.

Another option is to use a library such as `iframe-resizer` to solve this. Typically, this may look like the following snippet on the top frame:

```
<style>
  iframe {
    width: 1px;
    min-width: 100%;
  }
</style>
<iframe class="component" src="/iframe.html"></iframe>
<script src="/iframeResizer.min.js"></script>
<script>
iFrameResize({ log: true }, 'iframe.component')
</script>
```

And it may look like the following snippet on the actual `<iframe>` (for example, `iframe.html`):

```
<style>
  html, body {
    padding: 0;
    margin: 0;
  }
</style>
<!-- Content of the iframe -->
<script src="/iframeResizer.contentWindow.min.js"></script>
```

The main advantage of such an approach is that the layout will be dynamically optimized. This also works when the content changes. The drawback is that we need another script and therefore it reduces the performance of the website.

Summary

In this chapter, you learned about your first pattern to realize large-scale micro frontend solutions. Using a web approach, you can now go ahead and actually fuse different, more fine-grained web servers together.

On the outside, this feels like a single web server. The techniques are also not very exotic. By using standard proxies, links, and `<iframe>` elements, we can create a quite sound solution in no time.

The main advantage of the web approach is also its biggest drawback. The system is so simple that it can be composed arbitrarily. Therefore, it is quite difficult to ensure a working system.

In the next chapter, we will look at a more complicated enhancement of this pattern in the form of server-side composition.

7

Server-Side Composition

In the previous chapter, you saw that micro frontends can be composed with very simple methods. The web already gives us everything we need. The problem, however, with such simple solutions is that they don't scale well – neither in development nor at runtime. And scaling is one of the most important points about micro frontends.

Now, the question is: what can we do about this? How can we keep everything as isolated, separated, and independent as possible, without having any constraints on scalability? One possible answer can be found in the server-side composition pattern. It uses techniques available for web servers to join the micro frontends before they reach the client.

The trade-off of this pattern is that additional complexity needs to be introduced in the backend. As a result, we can dynamically stitch a view together from various sources using server-side composition. It allows horizontal micro frontends without requiring any shared repository at all. This chapter goes into all the details of how and when this pattern should be applied.

We'll start with some of the basics about this architecture pattern. Here, we will also introduce a sample implementation – this time using multiple repositories to fully see how this will work.

Afterward, we'll discuss the advantages and disadvantages of this pattern. Finally, we will go into two specific components of this pattern: what layouts are and how micro frontends will be developed in practice using this pattern.

We will briefly cover the following topics in this chapter:

- Basics of server-side composition
- Advantages and disadvantages of server-side composition
- Creating a composition layout
- Setting up micro frontend projects
- Using a dedicated rendering server

Let's jump right into the topics.

Technical requirements

To follow the code samples in this book, you need knowledge of JavaScript and how to use the command line. You should have Node.js (version 20 or higher) installed using the instructions at `https://nodejs.org`. For this chapter, additional knowledge in the Node.js web application framework Express can be helpful.

The code used in this chapter can be found at the following repo:

`https://github.com/PacktPublishing/The-Art-of-Micro-Frontends-Second-Edition/tree/main/Chapter07`

The CiA video for this chapter can be found at `https://packt.link/do1Xr`

Basics of server-side composition

A server-side composition implementing micro frontends requires a central point in the backend where the micro frontends are resolved and merged. The basic principle is shown in the following diagram:

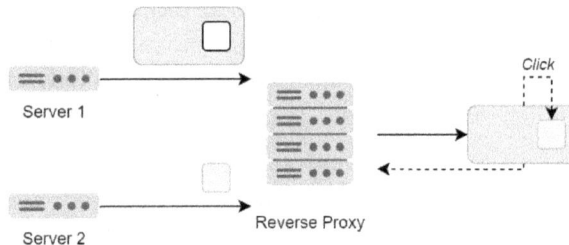

Figure 7.1 – The idea behind server-side composition – a central
server joins the different frontend fragments

The idea outlined in *Figure 7.1* is that the communication from the client is never directly targeted to the micro frontends. Instead, a reverse proxy is acting as a man-in-the-middle – forwarding the requests to the correct micro frontends, as well as merging the responses together to form single HTML pages that can be directly evaluated in the client.

Using this pattern, teams will either go for full pages or just individual fragments. Fragments can be as large as fully composed parts of a page or as small as UI components with some domain logic.

As mentioned, the reverse proxy essentially plays the same role as the aggregation layer introduced in the previous chapter. The difference is that its responsibilities go beyond proxying micro frontends. Therefore, we'll usually refer to this point as a gateway service or a **BFF**.

A BFF is a part of the backend that has been introduced only to serve the frontend. Generally, a BFF can be thought of as a layer between UX and required resources. In the preceding diagram, each server may be fully independent, however, only the BFF is able to join the individual resources to form one coherent UX, which is then served as a page.

The architecture

In general, the servers developed in the web approach could be used here, too. After all, these should just be serving some HTML fragments. The actual new piece is the aggregation layer, which has to stitch the different fragments together to form fully rendered pages.

As such links may still be used to navigate between pages, these pages are not necessarily isolated. Instead, pages are always composed at the aggregation layer. If by coincidence, a page is already fully provided by one micro frontend, then this page must still get through the aggregation layer where it will be processed before sending it to the client.

The aggregation layer has multiple responsibilities. The most important task is to resolve a template associated with the current page identified by its URL. Then the template needs to be expanded by proxying requests to the involved micro frontends. Finally, recursive resolutions and resource path adoptions need to be considered.

Let's use server-side composition for our existing sample project.

Sample implementation

For this sample, we'll go beyond a single repository. This way, we can see how micro frontends would work in real projects without having an artificial escape hatch. Furthermore, we'll use a more elaborate example application. We'll choose the famous tractor store from micro-frontends. org. Composed together, the page looks like this:

Figure 7.2 – The tractor store (v1) sample as shown on the official website

The tractor store (v1) represents a page composed of three micro frontends. One micro frontend deals with the products (red), another one covers the store logic (blue), and the third one integrates recommendations for related products (green).

Let's have a look at how team red can implement its micro frontend.

Implementing the products page – the red micro frontend

The red micro frontend is responsible for the product information page. In our sample, this page will be the central driver for all the content, as it represents the only page.

We can use a Node.js Express server to serve the content. With the help of the ejs and nodesi packages, we integrate a templating mechanism. This allows us to define our page using the following HTML:

```
<link rel="stylesheet" href="./product-page.css">
<h1 id="store">The Model Store</h1>
<esi:include src="/mf-blue/basket-info?sku=<%= current.sku
  %>" />
<div id="image">
  <div>
    <img src="./images/<%= current.image %>" alt="<%=
    current.name %>" />
  </div>
</div>
<h2 id="name">
  <%= product.name %> <small><%= current.name %></small>
</h2>
<div id="options">
  <% product.variants.forEach(variant => { %>
    <a href="./product-page?sku=<%= variant.sku %>">
      <button class="<%= current.sku === variant.sku ?
        'active' : '' %>" type="button">
        <img src="./images/<%= variant.thumb %>" alt="<%=
          variant.name %>" />
      </button>
    </a>
  <% }); %>
</div>
<esi:include src="/mf-blue/buy-button?sku=<%= current.sku
  %>" />
<esi:include src="/mf-green/recommendations?sku=<%=
  current.sku %>" />
```

The parts where we use variables are indicated via the placeholders <%= ... %>. The other thing to note is the use of the esi:include tag for referring to fragments from other micro frontends. Later on, we will introduce **Edge-Side Includes** (**ESI**) in more detail. For now, we only need to know that this is a possible mechanism to indicate the insertion of code from another micro frontend. The src attribute indicates from where the desired part is retrieved.

The document above is a valid HTML fragment and renders well in the browser, too. The final format depends slightly on the used aggregation layer. In general, we should limit ourselves to fragments as seen, even though the ideal case would be to write complete and valid HTML documents.

The Model Store

Tractor Porsche-Diesel Master 419

Figure 7.3 – The product page from the red micro frontend running in isolation

When rendered, the micro frontend already looks similar to the full page shown in *Figure 7.2*. Since the page's styling as well as the other micro frontends are missing at this point in time, we can certainly live with its plain look.

The boilerplate for the Express server is pretty much the same as the one we used in the previous examples. We use the following:

```
const express = require('express');
const path = require('path');
const { renderFile } = require('ejs');

const app = express();
```

```
const port = process.env.PORT || 2001;
const host = `http://localhost:${port}`;

// define the views and view engine
app.set('views', path.resolve(__dirname, '..', 'views'));
app.engine('html', renderFile);
app.set('view engine', 'html');

// define the folder for the assets
app.use(express.static('public'));

// start the server
app.listen(port, () => {
  console.log(`[OK] MF-Red running at ${host} ...`);
});
```

Additionally, for the product information of the tractor store, we'll need some data such as the available models and their properties. While this data might come from some API or ultimately a database in a real-world scenario, we'll just hardcode it for the example. Using the hardcoded data we can define the /product-page route, which represents the product information micro frontend:

```
app.get('/product-page', (req, res) => {
  const sku = req.query.sku || 'porsche';
  const current =
    product.variants.find((v) => v.sku === sku) ||
      product.variants[0];

  res.render('product-page', {
    product,
    current,
  });
});
```

This endpoint gets the information about which product should be used from the query parameter. Alternatively, we can make the SKU information part of the path. As the default, we fall back to some data point. Finally, we render the page using the product-page template with the two variables product and current.

The product page itself is interesting, but without the help from the blue micro frontend there is no possibility of any interaction. Let's implement this one, too.

The code for this section can be found at:

https://github.com/PacktPublishing/The-Art-of-Micro-Frontends-Second-Edition/tree/main/Chapter07/07-server-side-composition/red

Implementing store functionality – the blue micro frontend

Even though this micro frontend could be created using a completely different technology (for example, PHP), we'll just stick to the previous stack. In the end, our goal is not to show as much technology as possible, but to illustrate the pattern itself.

The blue micro frontend covers the management of the order basket. In the tractor store, this is split into two components:

- Basket info to display the number of items in the basket
- An **Add to cart** button allowing additional items to be added to the basket indicated in the basket info

The tricky part here is that this management involves a user-specific state. There are multiple ways to implement such a state in the backend, but most of these ways rely on a cookie to be transported via the headers.

For simplified session management, we'll use the `express-session` package. It gives us a straightforward wrapper around cookies and the management of user state. In this example, a simple in-memory store to manage the user state is sufficient.

Setting up the session integration works with the following code:

```
const session = require('express-session');

app.use(
  session({
    secret: process.env.STORE_SECRET || 'foobar-blue',
    resave: false,
    saveUninitialized: true,
  })
);
```

Since we will be needing form transmission to add items to the basket, we should also set up middleware to parse form data. Here, the easiest way is to use what is already given by Express:

```
app.use(express.urlencoded({ extended: true }));
```

Finally, to actually leverage the session cache, we can use the `session` property on the request object.

As an example, the following endpoint would render the basket info:

```
app.get('/basket-info', (req, res) => {
  res.render('basket-info', {
    count: req.session.count || 0,
  });
});
```

If no count was defined yet, we just set it to its starting value, which naturally is zero.

The basket-info template is then defined via the templating rules from the ejs package again:

```
<link rel="stylesheet" href="./basket-info.css">
<div class="blue-basket" id="basket">
  <div class="<%= count === 0 ? 'empty' : 'filled' %>">
    basket: <%= count %> item(s)
  </div>
</div>
```

Again, this is pretty much an isolated fragment that can also just render fine standalone. In the browser, this looks as follows:

Figure 7.4 – The basket info from the blue micro frontend running in isolation

With the functionality from the blue micro frontend, the overall application would already be fully functional. However, for a successful business, a product page will need a bit more – a way to discover other products that may be worth a look.

The green micro frontend covers this business aspect by bringing in a component with product recommendations.

The code for this section can be found at:

https://github.com/PacktPublishing/The-Art-of-Micro-Frontends-Second-Edition/tree/main/Chapter07/07-server-side-composition/blue

Implementing the list of recommendations – the green micro frontend

From the implementation point of view, the green micro frontend is definitely the simplest one. It provides one HTML fragment that represents the list of recommended products.

The fragment is also using the template language from the `ejs` package. It looks as follows:

```
<link rel="stylesheet" href="./recommendations.css">
<div class="green-recos" id="reco">
  <h3>Related Products</h3>
  <% recommendations.forEach(recommendation => { %>
    <img src="./images/reco_<%= recommendation %>.jpg"
      alt="Recommendation <%= recommendation %>">
  <% }); %>
</div>
```

Again, we isolate the styles for this view in a dedicated CSS file. By looping over the recommendations, we generate a list of images. These images will be retrieved by convention.

When the green micro frontend renders in the browser, it looks as shown here:

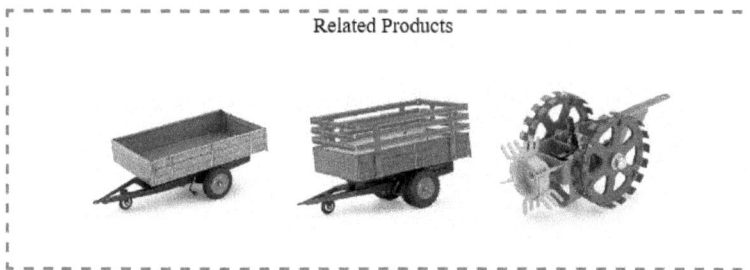

Figure 7.5 – The recommended products from the green micro frontend running in isolation

The endpoint is quite straightforward, too:

```
app.get('/recommendations', (req, res) => {
  const sku = req.query.sku || 'porsche';

  res.render('recommendations', {
    recommendations: allRecommendations[sku] ||
      allRecommendations.porsche,
  });
});
```

From its code base, there is nothing unique about the green micro frontend. Nevertheless, it shows again that a powerful templating language and resource-loading capabilities are necessary to ship server-side composed micro frontends.

With all the parts ready, it's now time to bring them together using the aggregation layer.

The code for this section can be found at:

https://github.com/PacktPublishing/The-Art-of-Micro-Frontends-Second-Edition/tree/main/Chapter07/07-server-side-composition/green

Implementing the gateway

We finish with the most important component of the described architecture – the aggregation layer. Again, we use Node.js with Express to create a simple yet powerful web server.

As mentioned, the aggregation layer comes with a layout – essentially a basic template to define the HTML document as seen by users. In the case of the tractor store, the layout could be defined like this:

```
<!DOCTYPE html>
<html lang="en">
  <head>
    <meta charset="UTF-8" />
    <title>Tractor Store</title>
    <link href="/page.css" rel="stylesheet">
  </head>
  <body>
    <div id="app"><esi:include src="<%= page %>" /></div>
  </body>
</html>
```

It's a full HTML document but contains a placeholder for the main content. The source of this content needs to be determined by the backend.

Before diving into the code of this server, let's recap what requirements for the gateway we need to fulfill:

- Proxying requests to micro frontends
- Replacing placeholder tags with contents from micro frontends
- Adjusting hyper references (for example, for links, style sheets, images, forms, and so on)
- Forwarding and aggregating cookies
- Serving static content

For simplicity, we'll use a static set of micro frontend targets:

```
const targets = {
  '/mf-red': 'http://localhost:2001',
  '/mf-blue': 'http://localhost:2002',
  '/mf-green': 'http://localhost:2003',
};
```

The configuration for the reverse proxy is quite straightforward too. We just need to configure the base URL and restrict the allowed hosts to the defined micro frontend targets. To avoid long-running requests due to cyclic references, we limit the maximum depth of placeholder replacements.

The last thing to consider in the reverse proxy settings is that caching may not be as straightforward as it seems. While caching is definitely a must, we should choose the configuration wisely. If we update parts of the content, we don't want to end up in an inconsistent state. For the sample, the easiest way around these fine-tuning issues is to disable caching altogether:

```
const esiConfig = {
  baseUrl: host,
  allowedHosts: Object.keys(targets).reduce(
    (prev, prefix) => [...prev, targets[prefix]],
    [host]
  ),
  maxDepth: 8,
  cache: false,
};
```

Any route prefixed with /page/ will resolve the default template, where the content comes from the micro frontend denoted in the path:

```
app.get('/page/*', (req, res) => {
  req.esiOptions = {
    headers: { cookie: req.headers.cookie },
  };
  res.render('default', {
    page: makeUrl(req.path.substr(5), req.query),
  });
});
```

For instance, requesting a page using the /page/mf-red/product-page path will render the template resolving the content from /mf-red/product-page, which will then proxy the request to the red micro frontend.

The proxy requests are resilient and should never crash the application. We can test this by turning off individual micro frontends. As an example, if we turn off the red micro frontend, only the plain template without any content is rendered.

Figure 7.6 – The gateway running in isolation without any micro frontends

The most difficult parts are the successful submission of forms and the actual adjustment of hyper references. For form submissions, we can come up with a solution that forwards the form's body to the target micro frontend and discards the response.

Ideally, we would reuse the previous HTML code and replace the part corresponding to the micro frontend submitting the form. However, the problem with this approach is that a form submission may influence parts from other micro frontends, too. That way we would need to dynamically know which ones can be kept and which ones would need to be replaced. We are back at our initial caching problem.

As a quick fix, we'll never actually care about the response to form submissions in our quick sample. While full frameworks will be more sophisticated in that respect, we only need to know the basics.

Making a form request can be done easily with a library such as `axios`. Since this may involve state, we should inject the cookie of the current request – and retrieve the potentially modified cookie to send it with the response later on. In code, this can be done by setting the `cookie` header explicitly:

```
axios.request({
  method: 'POST',
  data: req.body,
  url: target,
  headers: { cookie },
})
.then(({ headers }) => res.header('cookie',
  headers.cookie))
.finally (() => res.redirect(req.headers.referer));
```

Our final rendering needs to be the same as the page we came for. For this, we can use the `referer` header. By redirecting to it, we'll make sure to reuse the existing infrastructure for page rendering, too.

The other difficult part was the hyper reference adjustment. The difficulty here is that we need to have some HTML parsing skills, otherwise we may get the wrong URLs or elements. Just making a lookup via some regular expression may work in simple cases, but avoids edge cases where the actual DOM or source code details such as comments play an important role.

The least complex way to implement this is to inspect the response of a proxy request. If the response contains HTML, then a parser such as `cheerio` may load and inspect the delivered HTML.

We'd go over all relevant elements, such as `<img>` to look at hyper references that need to be adjusted. The general rule could be to only adjust relative URLs that do not start with one of the known target prefixes (for example, `/mf-red`). Finally, we serialize the parsed and manipulated DOM back to HTML again, which is then used as a fragment in the final document delivered to the client.

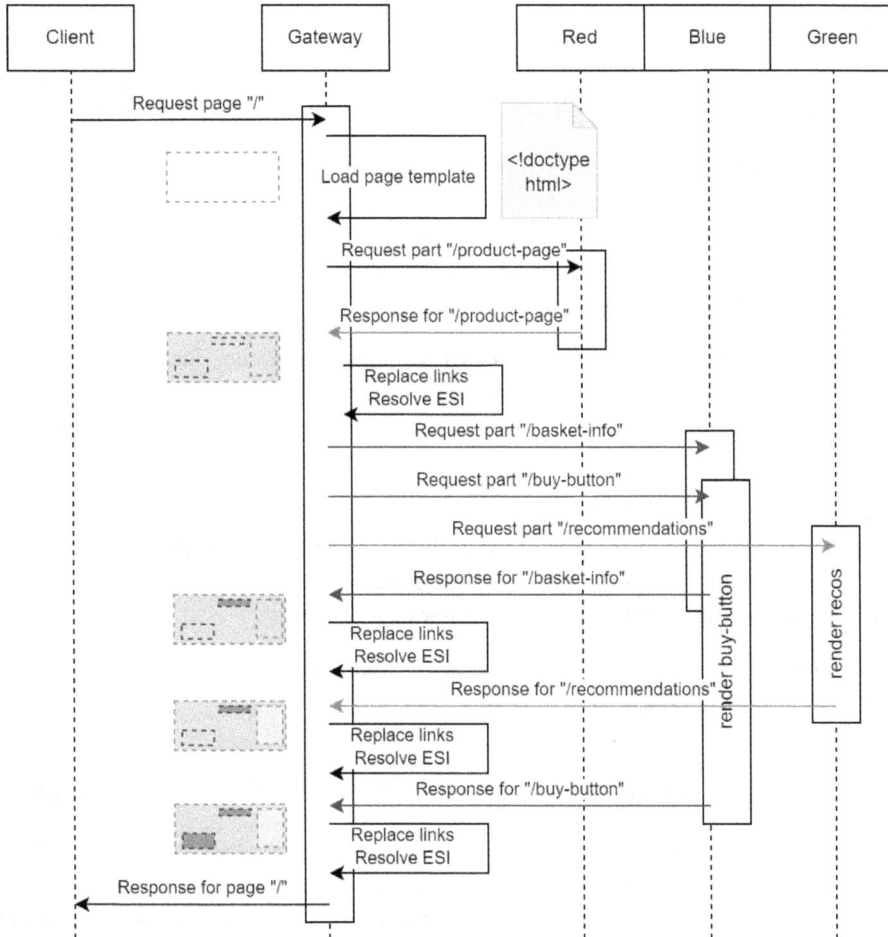

Figure 7.7 – Sequence diagram showing the different calls when requesting a page from the gateway

The whole flow is shown as a sequence diagram in the preceding illustration. Note that this sample only exercises two levels in the fragment resolution, one in the template coming from the gateway and the other in the response from the red micro frontend. In reality it could be much deeper, potentially requiring a recursion limit or some flattening to avoid long response times.

Having finished this sample, the question is what could we have done better? Let's find out.

The code for this section can be found at:

```
https://github.com/PacktPublishing/The-Art-of-Micro-Frontends-Second-
Edition/tree/main/Chapter07/07-server-side-composition/gateway
```

Potential enhancements

Like in the last pattern of *Chapter 6*, we should introduce a linking directory to decouple the micro frontends from the aggregation layer. This will also enable the scaling behavior we desire.

Another thing to consider is a mechanism to store the different templates. Together with the definition of a template format, this forms the heart of the micro frontend solution. In the end, server-side composition can be thought of as a kind of CMS where the individual building blocks are highly functional and pre-rendered by different services. It is no surprise that this pattern is most useful for information-heavy websites.

The resolution of the contained fragments needs to be more sophisticated too. Here, we could introduce maximum response times and retries. We could also set caching rules to identify what we can safely cache and what we cannot. Finally, we can configure form substitutions to reduce the number of page requests.

Since we talk about similarities with CMS applications, what we could introduce is a fragment store, which allows the pre-rendering of any part of the available micro frontends. This can be thought of as an enhanced kitchen sink for these components; something like Storybook for micro frontends. This helps to discover existing pieces and compose new pages faster.

But even without these enhancements, the server-side composition may be quite appealing. The advantages and disadvantages speak for themselves.

Advantages and disadvantages of server-side composition

Server-side composition is among the most used patterns for implementing micro frontends. There are several reasons for this. On one hand, we have a pattern that scales well, encourages loose coupling, and has great performance. On the other hand, this pattern allows dynamic discovery, independent development, and a high degree of flexibility.

As usual, there are a couple of disadvantages, too. Since different frontends are merged into a single page, there is no way to ensure proper isolation. Consequently, scripts and style sheets may conflict with each other. Conflicts may also appear more often due to a lack of proper debugging tools.

While debugging an individual micro frontend is pretty much as straightforward as for the web approach, debugging a composed website is quite difficult. Getting a proper local development flow for this pattern can be quite complicated. This is the time when established frameworks such as Mosaic 9 or Podium come in handy.

Reviewing Mosaic 9

One of the reasons why local development of the full solution can be complicated is that server-side composition solutions usually combine multiple services and technologies to be operational. As an example, Zalando's open source solution **Mosaic 9** consisted of the following parts:

- **Tailor**: A layout service
- **Skipper**: An extensible HTTP router
- **Shaker**: Provides fragments to be available to all micro frontends
- **Quilt**: Template storage used by Tailor
- **Innkeeper**: A linking directory used by Skipper
- **Tessellate**: A service to render React components

As mentioned, all these services need to be running locally to enable full local development. For Mosaic specifically, we would have needed that setup for individual micro frontends, too. Otherwise, the full spectrum of possibilities can either only be used or seen online.

While Mosaic 9 may be an extreme example, others try to stay more compact and lightweight. An example here is Podium.

Introducing Podium

Podium tries to pretty much do the crucial infrastructure in a single reusable service written in Node. js. While the framework itself claims to be technology-agnostic, it comes with an implementation for the Express framework. There are plugins for Hapi and Fastify, too.

> **Important note**
>
> Node.js has a couple of HTTP frameworks. Some are thin; others come with more integrated functionality and are consequently more opinionated. Express is arguably the most used framework, however, since it's also among the oldest ones, it misses some of the more modern patterns and convenience methods. Frameworks such as NestJS, Nitro, Hapi, or Fastify try to be more modern alternatives. All of these (and many more) are totally capable of doing the job. Make sure you align with the team when choosing one over the others.

There are only two concepts in Podium:

- A layout service to merge micro frontends into rendered pages
- A **podlet**, which is an application serving a micro frontend

The problem with the given simplicity – especially in contrast to Mosaic 9 – is that not all necessary areas are covered. For instance, while Mosaic 9 comes with services to manage a linking directory or to render React components, none of that is available in Podium. This does not really matter if the project will be static or will not use React, but otherwise, such functionality must be added later on.

Nevertheless, especially for starting server-side composition, Podium is a great fit. The local development process is much easier to set up than in many frameworks. The requirements are quite minimalistic, which reduces the necessary infrastructure and saves on costs. Essentially, it all boils down to running the layout service, which could be configured to resolve against podlets running online or in the local environment. Since Podium is missing a linking directory, how to do this exactly is up to the team using it.

More information on Podium can be found at `https://podium-lib.io/`.

Known users

Zalando was among the first to advocate this pattern, and they started refining and iterating on it soon after. However, in the meantime, many companies joined. One industry where this pattern is especially strong is e-commerce. Here, companies such as the German brand Otto, or the well-known global-player IKEA put their faith in micro frontends.

The reason for the success of server-side composition in online shops can be explained by looking at the advantages. In this pattern, performance is king. Also, since the composition happens without any requirement on the client, the broadest possible audience can be reached. In the end, this pattern has the least possible negative impact on customer behavior, which relates directly to turnover.

Knowing this one key fact makes you wonder: why not stop here? Why introduce more patterns when server-side composition is already quite successful? Well, there are things that don't play well in server-side composition. One of these is the single-page application style of building an application.

Creating a composition layout

As already mentioned in the basics of server-side composition, server-side composition always uses a gateway service to merge the different micro frontends. Even though this will be a reverse proxy in most cases, usually this does a bit more than just proxying the response. Actually, the reverse proxy part is just one aspect of it. The more important one is the layout.

In the following sections, we'll look at its responsibilities and some techniques to include fragments including SSI, ESI, and JS template strings. We'll start with its responsibilities.

Understanding layout responsibilities

A layout is responsible for structuring a page. It determines where all the individual fragments from the different micro frontends should be placed. There are two kinds of layouts:

- Generic layouts
- Specific layouts

A generic layout is determined by the gateway service. This could either be hardcoded or resolved by some logic, for example, from another service or database. In contrast, specific layouts are determined by the individual micro frontends. These two concepts are not exclusive.

For instance, a generic layout could determine the header, footer, navigation bar, and content. On the other hand, the content may be resolved exclusively by a specific micro frontend, which then uses another layout to mix between its content and another sidebar coming from some other micro frontend.

The only requirement is that only the gateway service is responsible for properly resolving the layout to produce some HTML that can be sent back to the client.

For this reason, the gateway service is sometimes called a layout service or layout engine. Of course, as seen in the Mosaic 9 case, this may be put into another microservice, too, to keep the gateway as lightweight as possible.

To introduce the structuring aspect, quite a variety of technologies can be used. As mentioned, server-side includes may be sufficient in some cases. In most cases, however, more advanced techniques such as ESI are better. Presumably the most powerful, yet least flexible, is to hardcode the layout by using a template defined with a programming language, for example, using JS template strings.

Using SSI

The beauty of SSI is that the technique is so old and established that it works or can work for almost all web servers. Even if it's not supported out of the box, there is a high chance that a plugin or extension exists. Otherwise, implementing SSI is also not too difficult.

The other beautiful property of SSI is its progressive nature. Even if we were to use a layout utilizing SSI on a web server without SSI support, it would not do any damage. Instead, nothing would be rendered. The main reason for this is the choice of an HTML comment as a carrier in the HTML source code:

```
<!--#include virtual="footer" -->
```

Besides some common directives such as `include` or `echo`, the SSI standard also defines control directives such as `if`. These allow using statements to conditionally render or hide parts of a layout.

Another great advantage of using SSI is that nothing besides actual HTML is required for defining this. Therefore, layouts may be stored in a database. One could use a CMS-like experience for actually creating and updating layouts, making them quite fast to iterate and refine.

Using ESI

In comparison to SSI, the **Edge-Side Includes** (**ESI**) standard looks much more modern. Instead of HTML comments, XML-based ESI tags are used. Such a tag could look like this:

```
<esi:include
    src="/footer"
    alt="/empty"
    onerror="continue" />
```

Besides a much richer set of instructions and a lot more scripting functionalities, the biggest advantage over SSI is the introduction of error handling. This allows you to use a failover in the case of an unavailable server.

The disadvantage is that ESI is more complex to implement. Usually, specialized reverse proxies such as Varnish or nginx are standard web servers considered when using ESI. Their configuration may not be easy and could be too limited. An alternative in the Node.js ecosystem is the `nodesi` package that we used beforehand in the sample implementation. It comes with direct support for the Express framework and supports a subset of the full ESI specification.

> **Important note**
>
> **Varnish** is often referred to as an HTTP accelerator, which is used especially in content-heavy websites. It comes with a strong caching system that works particularly well with dynamic websites and APIs. Additionally, load balancing, compression, streaming, plugins, and scripting via **Varnish Configuration Language** (**VCL**) are supported. VCL is transpiled into C and then compiled to ensure the fastest possible execution. More information can be found at `https://varnish-cache.org/`.

In general, the advantages of SSI are kept with ESI. Likewise, we could store layouts as simple HTML in a database. This would invite non-developers to create and update layouts. Not only would it increase the team size, but it would also allow closer collaboration with marketing specialists, UX designers, and technically strong product owners.

Sometimes, however, these markup extensions lack some of the power that developers are either used to or need to use. In such cases, we can always join the micro frontends programmatically.

Using JS template strings

Using a template string is an easy way of defining a string that's quite flexible in any programming language. In JavaScript, the ES6 standard introduced template strings, which allow composition using expressions enclosed by $ { }. As an example, consider the following code:

```
const name = 'Florian';
const age = 39;
```

```
const content = `My name is ${name} and next year I'll be
  ${age + 1} years old.`;
```

If we get all necessary fragments and store them in dedicated variables, we can join them programmatically using a template string. Now it's only a matter of knowing which fragments to retrieve and actually fetching these individual parts.

For instance, using Podium the different parts need to be fetched manually, too. Nevertheless, the framework still helps to keep the code quite minimalistic.

Here's one simple example using a layout combining a page from two fragments, namely `content` and `navigation`:

```
const Layout = require('@podium/layout');
const app = require('express')();

const layout = new Layout({
  name: 'homePage',
  pathname: '/home',
});
const navigationClient = layout.client.register({
  name: 'navigation',
  uri: 'http://localhost:7001/manifest.json',
});
const contentClient = layout.client.register({
  name: 'content',
  uri: 'http://localhost:7002/manifest.json',
});

app.use(layout.pathname(), layout.middleware());

app.get(layout.pathname(), async (req, res) => {
  const page = res.locals.podium;

  const [navigation, content] = await Promise.all([
    navigationClient.fetch(page),
    contentClient.fetch(page),
  ]);

  page.view.title = 'Home';

  res.podiumSend(`
    <nav>${navigation}</nav>
    <main>${content}</main>
```

```
    `);
});
app.listen(7000);
```

The downside of a programmatic approach using string templates is that updates can only be made by updating the full service. This will prevent rapid iterations and fast enhancements. Instead, layouts are essentially code, which needs to be deployed and maintained by developers.

The other thing that developers will need to maintain anyway is the individual micro frontends.

Setting up micro frontend projects

Micro frontend projects are just independent web servers that can be reached from the aggregation layer. There are various frameworks and tools that try to make micro frontend development for server-side composition as simple and straightforward as possible.

There are three potential ways to simplify micro frontend development:

- Using a serverless approach where the whole runtime is already given

- Providing a scaffolding tool to create project boilerplates

- Having a sample that can be cloned and adjusted

In general, these three options are not exclusive. It is possible to use a serverless approach, which comes with a project scaffolding option and has some examples available to illustrate how development works.

Serverless approach

In recent years the **Function-as-a-Service (FaaS)** approach has boosted service development. Offerings such as AWS Lambda or Azure Functions allow us to write a full service by just providing a handler function, which is then called from a runtime within the respective platform. This gives us great scalability with the least maintenance effort. FaaS can also be implemented internally, where teams then only provide handler functions without having to deploy full services that include a runtime.

In the case of Podium, the development of a micro frontend is boosted by using the existing @podium/ podlet Node.js package. Keep in mind that Podium tries to be framework-agnostic and could also be used without this package.

As the name of the Node.js package already suggests, Podium names its micro frontends podlets. Let's see how to create one of these podlets using Node.js and their helper package.

Podlets

In an illustrative scenario, everything necessary for a working podlet fits in one file, representing a full Node.js web server. We start with the imports, getting the helper package and Express instantiated:

```
const Podlet = require('@podium/podlet');
const app = require('express')();
```

Now we can define the metadata of the current micro frontend. Metadata will play a more important role in the upcoming patterns. But even at this point, we should properly expose our micro frontends, at least communicating their name and version.

In the given example, we create a micro frontend to represent the content on the home page:

```
const podlet = new Podlet({
  name: 'homeContent',
  version: '1.0.0',
  development: false,
});
```

At this point in time, we can set up all the routes of our web server. This is also where our micro frontend logic will need to appear.

In Podium, we need to have an endpoint that serves the podlet's metadata. Finally, we start the web server at a defined port:

```
app.use(podlet.middleware());
app.get('/manifest.json', (req, res) => {
  res.json(podlet);
});
app.get('/', (req, res) => {
  res.podiumSend(`<section>Welcome!</section>`);
});

app.listen(7002);
```

In this example, we use a different port than for the other micro frontends. This, however, is quite artificial. Usually, we'd use the same port (for example, 8000) and either use unique ports in our final setup with containerization or via a given configuration.

Can this podlet run independently? Sure. After all, this is just a standard Express application. There is no communication or configuration here that leads to the template service. How does a development lifecycle then look for server-side composition using Podium as an example? We will get the answer to this question in the next section.

Examining the lifecycle

If we use Podium without any extra services, we'll end up with a rather static setup. All the micro frontends need to be known statically in the template service. As such, we'll end up with three cases:

- The creation of a new podlet
- Updating an existing podlet
- The removal of an existing podlet

In the first and the last case, we'll need to update the template service when we publish the change. In the second case, the team responsible for the podlet can indeed work independently. Depending on the project setup, the independent workflow may allow everything from making a pull request to accepting it, to publishing the update to the service. Some workflows are more restrictive and require the explicit authorization of a central team.

This leaves us with only one path where a central team needs to get active. The solution would be a linking directory available on a central location where new services would be automatically discovered and included.

Finally, before we can move on to the next pattern, we look at a variation of the server-side composition that uses a single rendering server instead of multiple.

Using a dedicated rendering server

The key idea behind the server-side composition is to have a central template server that works in combination with multiple rendering servers. Consequently, this is essentially producing a composition of multiple independent rendering servers into a single website. A natural question to ask is why we need multiple servers and what advantages and disadvantages we would have if only a single server is used.

Starting with the basics a single rendering server is a lot less infrastructure to operate – and pay for. With the rendering process in the hands of a single server, we also get a lot less internal network traffic. Finally, sharing state or optimizing rendering output is a lot simpler than the HTML manipulation we've had to deal with earlier in this chapter.

On the downside, the freedom of technology for micro frontends will be drastically reduced in most cases. While a certain degree is still possible, a lot of constraints coming from the rendering process defined on the server will be naturally imposed on the micro frontends.

As an example, if we decide that the central rendering server only works with React in a streaming mode then every micro frontend needs to be written using JavaScript (or TypeScript) with React. Other choices are not allowed. While support for other JavaScript-based frameworks could be added with medium effort, frameworks from other technologies such as ASP.NET Core or Spring Boot will be either quite difficult or simply impossible to integrate.

One thing that is surprisingly not a disadvantage is performance. In this case, single-server does not imply single-instance – therefore the scaling behavior of the rendering server is as good as beforehand. If we need more performance, we can always just increase the server's resources or run another instance of the server behind a load balancer.

In *Figure 7.8*, the general idea behind this architectural variation is shown. The fragments are all rendered on the rendering server while the teams can still provide additional data using their own servers.

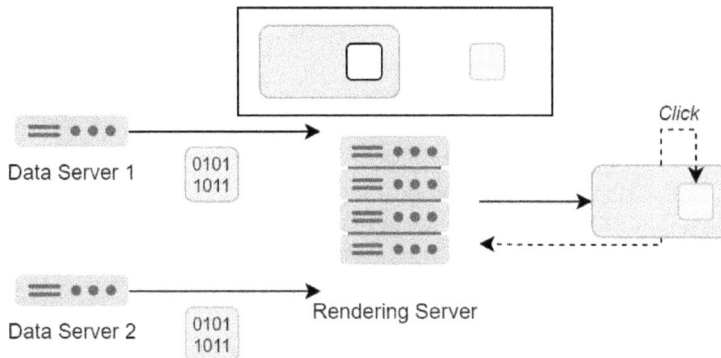

Figure 7.8 – The idea behind a dedicated rendering server – fragments
are rendered on a single server with data still distributed

One interesting side-effect of this variation is that it allows to create a special composition referred to as the **islands composition**. In the islands composition, a single rendering server is responsible for creating a full website, which is then brought to the user's browser with additional information regarding the deferred loading of JavaScript files depending on the usage of the website. Without micro frontends, the concept of having smaller parts of the website requiring the on-demand loading of JavaScript files is referred to as **islands of interactivity**.

> **Islands of interactivity**
>
> The island architecture was been introduced by Jason Miller based on the work of Katie Saylor-Miller. This architecture encourages the use of small, focused chunks of interactivity within server-rendered websites. The output is standard HTML with an orchestrator that allows progressive enhancement during the website's lifetime in the user's browser. In contrast to SPAs, there is no single-entry point – every island is an independent component that gets JavaScript enhanced under certain conditions such as when the area of the website where the component is shown is visible. You can read more about it at `https://www.patterns.dev/vanilla/islands-architecture/`.

The single rendering server makes it possible to still have islands of interactivity while obtaining parts of the website from different teams. This way, we can obtain optimal website performance while still being able to distribute work efficiently.

While it is possible to create a website using the islands composition pattern from scratch, we can also start with an implementation on top of an existing framework such as **Deno Fresh**, **Astro**, or **Qwik**. All of these implement the island architecture in JavaScript.

We will go into more details on islands composition later on with a bit more on hydration and progressive rendering in *Chapter 10*, as well as a primer on how to create dynamic web experiences with optimal performance using Qwik in *Chapter 11*.

Let's stop here and recap what we discussed in this chapter.

Summary

In this chapter, you learned how server-side composition can help us to bring together micro frontends already on the backend. Using server-side composition, we get the advantages of the web approach without much indirection and performance penalty. You've seen that many tools and frameworks exist to help us implement server-side composition swiftly, as well as a more complete example for implementing this pattern from scratch.

Server-side composition makes the most sense for information-driven web applications such as e-commerce websites, where fast response times and less usage of JavaScript are important. The complexity of the setup and the required infrastructure need to be considered before investing in the implementation of this pattern.

A potentially much simpler alternative is to use a single rendering server. This also has the advantage that data can still be distributed, but the rendering process is done on a single machine, leaving much more room for optimizations and delivery enhancements. This way, the perceived performance of a website can be optimized very well.

The information presented in this chapter should help you to decide in favor of or against using the server-side composition pattern. Following the code derived for implementing the tractor store sample, you can start to create large-scale micro frontend applications using server-side composition.

In the next chapter, we look at a more lightweight refinement of server-side composition in the form of edge-side composition.

8

Edge-Side Composition

In the previous chapter, you saw the complexity and potential gains that are involved when using server-side composition. As already mentioned in *Chapter 7*, a slightly simpler variant of server-side composition is the edge-side composition pattern. Besides being a less complex solution, this pattern aims to bring bigger performance improvements compared to server-side composition.

While edge-side composition usually lives on edge servers such as a CDN, the enhancements of this pattern can be applied on-premises, too. In the end, the ideas behind this simplification are what matter here. If the implementation is simple enough to be deployed on a CDN, then we know it fits this pattern.

One of the most important things we'll do in this chapter is simplify the example from the previous chapter. For this reason alone, it makes sense to have a closer look at edge-side composition. Again, for didactic purposes, we will not use a full-fledged micro frontend framework. Instead, we will work with standard tools that most web developers have used already.

We'll cover the following topics in this chapter:

- Basics of edge-side composition
- Advantages and disadvantages of edge-side composition
- SSI and ESI
- Stitching in BFFs

All in all, we will look at some of the topics we introduced in the previous chapter in more detail, but this time, from a usability and simplification point of view.

Without further ado, let's jump right into the topics!

Technical requirements

To follow the code samples in this book, you need knowledge of JavaScript and how to use the command line. You should have Node.js (version 20 or higher) installed using the instructions at `https://nodejs.org`. For this chapter, additional knowledge of Docker and nginx is required to understand all code samples.

The code used in this chapter can be found at the following repo:

`https://github.com/PacktPublishing/The-Art-of-Micro-Frontends-Second-Edition/tree/main/Chapter08`

The CiA video for this chapter can be found at `https://packt.link/8S1gP`

Basics of edge-side composition

Edge-side composition may be the oldest pattern besides the web approach – if we reduce the web approach to purely taking links, that is. As you already know, techniques such as SSI and, later, ESI have only been invented to place fragments of HTML on an HTML page. This was a simple yet flexible way to create reusable layouts.

Of course, when SSI was introduced, content delivery networks and distributed development weren't around. However, as we saw in the previous chapter, the use of SSI or its successor, ESI, maybe a great choice when you're looking for a good way of denoting UI insertion points. The best argument in favor of using SSI or ESI, in ESI's favor, is its widespread adoption and the clear rules governed by a specification.

As in the previous chapters, first, we will go over the architecture before doing a sample implementation. Finally, we will touch on some potential enhancements for our sample implementation.

The architecture

Unsurprisingly, the architecture diagram of the edge-side composition pattern does not look much different from the server-side composition pattern. As shown in the diagram of *Figure 8.1*, the crucial difference lies in the usage of the aggregation layer.

The challenge for the micro frontends in this architecture is that they must know about the rather dumb stitching mechanism. Due to this, some of the things that we introduced and worried about in the previous chapter are not a problem.

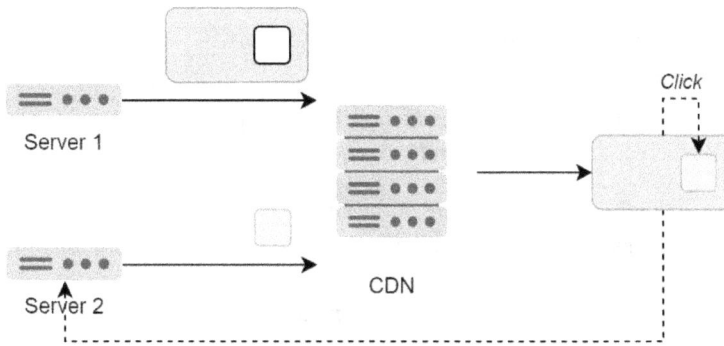

Figure 8.1 – Idea behind the edge-side composition pattern – fragments
are stitched together on the edge such as a CDN

For instance, submitting a form would work against the whole page. Instead of leaving this difficult problem to be figured out on the gateway, the micro frontend already knows how to solve this.

Depending on the edge-side offering that's used, all the aggregation layer can do is stitch together the different pieces in a rather static but cached HTML document, ready to be delivered to any user. There are other offerings that allow custom code to be run, too. Even using serverless functions may still be viable, if we wish to integrate micro frontends on the edge.

Let's go ahead and adjust our previous example so that it works on the edge, before deploying it to an existing CDN offering.

Sample implementation

In the previous chapter's example, we took the tractor store and made it work in a distributed manner using a very powerful aggregation layer. To follow the architecture of the edge-side composition pattern, we will need to reduce the responsibilities of this aggregation layer – otherwise, the aggregation layer cannot run on the edge infrastructure. Consequently, this means that each micro frontend needs to do more work and know more about the composition.

Realistically speaking, the backend-driven tractor store is a rather unfortunate example for edge-side composition. In *Chapter 9*, we'll see that this can be changed by placing some of its functionality on the client. Nevertheless, we can still make it work by dropping some of the usual benefits that can be obtained by using a CDN, such as caching.

For this example, we'll go back to a single repository. This choice is only due to simplicity and does not reflect any constraints on a repository's structure using edge-side composition. Starting with the sample from server-side composition, we only need to make the following changes:

- Replace the gateway with a more appropriate edge layer, such as a CDN offering, Varnish, Apache, or nginx – just to name a few.

- Modify the code from the red micro frontend to deliver the full page. This will be used as the basis for the edge layer.

- Modify the views from all the micro frontends to return the fully prepared URLs, which will resolve to the correct files or micro frontend services, respectively.

Our intention is to make the previous sample fit the edge-side composition pattern with the least number of changes. For instance, in the green micro frontend, we only need to adjust the paths. Instead of relying on micro frontend-independent paths such as /recommendations, we need to introduce the prefix again (for example, /green/recommendations). This should also be reflected in the view:

```
<link rel="stylesheet" href="/green/recommendations.css">
<div class="green-recos" id="reco">
  <h3>Related Products</h3>
  <% recommendations.forEach(recommendation => { %>
    <img src="/green/images/reco_<%= recommendation %>.jpg"
      alt="Recommendation <%= recommendation %>">
  <% }); %>
</div>
```

The previous convention of using relative paths that are then transformed on the gateway has now been replaced by absolute paths that contain the micro frontend-specific prefix. Otherwise, instead of introducing too many changes, we kept the ejs syntax such as <esi:include src="/mf-blue/basket-info" />.

In general, the blue micro frontend has also seen similar changes. The only difference is that this micro frontend also has a POST endpoint. Transforming this results in the following change:

```
app.post('/blue/buy-button', (req, res) => {
  // ... as before

  res.redirect(`http://localhost:1234/red/product-
    page?sku=${sku}`);
});
```

Now, we can redirect to the address of the edge layer. This assumes that we know the address of the edge layer.

Finally, for the red micro frontend, we need to introduce the URL changes and bring in the layout that was formerly kept as a template in the gateway layer. We also need to refer to all the other micro frontends. The way we refer to them now is totally edge-side dependent. If we were to use nginx, for example, we might end up using SSI as ESI is only supported via a plugin, while SSI is directly supported.

Making these changes gives us the following new product page content:

```
<!DOCTYPE html>
<html lang="en">
  <head>
    <meta charset="UTF-8" />
    <title>Tractor Store</title>
    <meta name="viewport" content="width=device-width,
        initial-scale=1.0">
    <meta http-equiv="X-UA-Compatible" content="ie=edge">
    <link href="/red/style.css" rel="stylesheet">
    <link href="/red/product-page.css" rel="stylesheet">
  </head>
  <body>
    <div id="app">
      <h1 id="store">The Model Store</h1>
      <!--# include virtual="/blue/basket-info?sku=<%=
        current.sku %>" -->
      ... as before
      <!--# include virtual="/blue/buy-button?sku=<%=
        current.sku %>" -->
      <!--# include virtual="/green/recommendations?sku=<%=
          current.sku %>" -->
    </div>
  </body>
</html>
```

Essentially, this is the result of pre-evaluating the previous template against the red micro frontend exclusively. Unfortunately, the red micro frontend needs to know the prefixes of the different sources it uses. Due to possible redirects and other tricks, this is not a red flag, even though an even looser mode would be appreciated.

Finally, on the edge layer, we will go for nginx. There are multiple reasons for this:

- nginx is a fairly easy-to-understand configuration
- It can be run within a Docker container without any problems
- It is the basis for many web applications already
- Its performance is great, making it a good choice for production, too

Our job is to configure nginx in such a way as to get the correct reverse proxy behavior. Here, we need to configure one section per prefix to redirect to the right server.

In our example, with the settings we used last time, we can come up with the following configuration:

```
server {
  listen 80;
  server_name frontend;

  location /red {
    ssi on;
    proxy_pass http://red:2001;
    proxy_set_header Host $http_host;
  }

  location /blue {
    ssi on;
    proxy_pass http://blue:2002;
    proxy_set_header Host $http_host;
  }

  location /green {
    ssi on;
    proxy_pass http://green:2003;
    proxy_set_header Host $http_host;
  }
}
```

We could set a lot more options here (such as custom headers to indicate the use in reverse lookups or caching behavior), but the preceding configuration is sufficient to get started.

The URLs we've used here need to be replaced by the actual URLs where these micro frontends run. In a local setup, where the micro frontends and the edge can be run in the same Docker network, special DNS names such as red, blue, or green can be used together with their respective ports to establish a connection.

The Dockerfile file for getting this configuration to run can be as simple as the following two lines:

```
FROM nginx:latest
COPY ./nginx.conf /etc/nginx/conf.d/default.conf
```

With these little changes, our solution is fully capable of working with an edge layer instead of a more resource-hungry aggregation layer within our own infrastructure.

To run the application locally you can create another `Dockerfile` for each of the micro frontends, i.e., each Node.js server running in another Docker container:

```
FROM node:slim
ENV NODE_ENV development
# Setting up the work directory
WORKDIR /express-docker
# Copying all the files in our project
COPY . .
# Installing dependencies
RUN npm install
# Starting the micro frontend application
CMD [ "node", "lib/index.js" ]
# Exposing server port, change accordingly
EXPOSE 2001
```

To combine the different Docker containers into a single network we can leverage **Docker Compose**. The definition of a Docker Compose network is done by writing a `docker-compose.yml` file.

In the given example the file could look like this:

```
services:
  web:
    build: ./edge
    ports:
      - "1234:80"
  blue:
    build: ./blue
  green:
    build: ./green
  red:
    build: ./red
```

This definition will create four containers running in the same network. Only the application running nginx on port 80 of the web container is reachable from the outside – accessible from port 1234. All other containers are only reachable inside the network.

For now, let's step back and look at the potential enhancements we can make to our demonstration sample.

Potential enhancements

In server-side composition, we have always been striving for loose coupling, for example, by introducing a dynamic linking directory. Since edge-side composition comes with a less powerful aggregation layer, this enhancement should be dropped… or should it?

It turns out that a linking directory may still be relevant if we use edge-side composition as an enhancement on top of server-side composition. In this scenario, edge-side composition resolves all micro frontends over a reverse proxy, which acts as the aggregation layer of server-side composition. The only difference is that this aggregation layer will only be responsible for resolving micro frontends and providing page templates. It will not directly stitch the micro frontends together.

Another enhancement is to allow cache invalidation from every micro frontend directly. While performing frequent updates from the CDN is acceptable, to begin with, it can easily create a lot of unnecessary loads and actually be undesired. In the end, a polling mechanism will always be unsuccessful and thus unnecessary or too late.

Besides these enhancements, what advantages and disadvantages are there in terms of using edge-side composition? Let's find out.

Advantages and disadvantages of edge-side composition

To cover the advantages and disadvantages of this pattern, we need to look at its close companion: server-side composition. If done correctly, then edge-side composition will be much more lightweight and allow us to do things such as caching. This gives websites a great performance boost, without us requiring more sophisticated algorithms or tricks.

One of the reasons why even a non-cached response may be faster with edge-side composition is that fragments are supposed to be flat due to the much-simplified aggregation layer. In the previous pattern, we were not only able to utilize nested fragments, but were actually encouraged to do exactly that. But even a non-flat structure can be flattened by using a simple trick. We'll illustrate this by using code.

Let's say we start with a structure like this:

```
// index.html (original)
<esi:include src="http://example.com/fragment1.html" />


// fragment1.html
<esi:include src="http://example.com/fragment2.html" />


// fragment2.html
<div>...
```

We can transform this into a flat structure by placing a pre-evaluated version of `fragment1.html` on the CDN:

```
// index.html
<esi:include src="./fragment1.html" />
// fragment1.html (original)
<esi:include src="http://example.com/fragment2.html" />
// fragment1.html (as served)
<div>...
```

So, instead of requesting `fragment-1.html` directly from the absolute URL, we are requesting it from the same CDN. This leads to us having a pre-evaluated version, which does not contain any ESI directives.

Still, the recommendation with edge-side composition would be to keep it as flat as possible – without having to perform any tricks. This will make debugging much simpler. Another area that benefits from this gained simplicity is caching.

Let's recap on the behavior we saw in the previous chapter when we implemented the sample. Requesting a page from our aggregation layer spawned a few requests.

The following diagram illustrates the time that was spent waiting for the response:

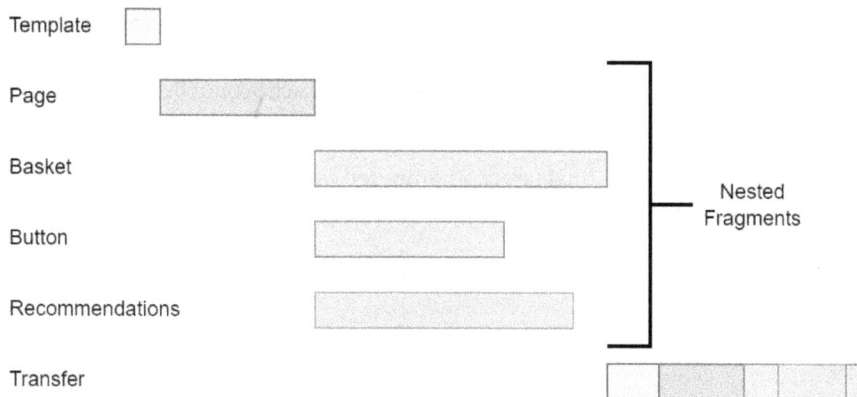

Figure 8.2 – The example of server-side composition, started with a template that led to a nested fragment

Following edge-side composition, we'll need to flatten this. Still, the fragment with the worst response time will dominate the page generation time, and therefore the overall response time. The transfer time will remain the same, though.

On the other hand, even considering the still apparent disadvantages, flattening has the most positive performance impact. The following diagram shows this:

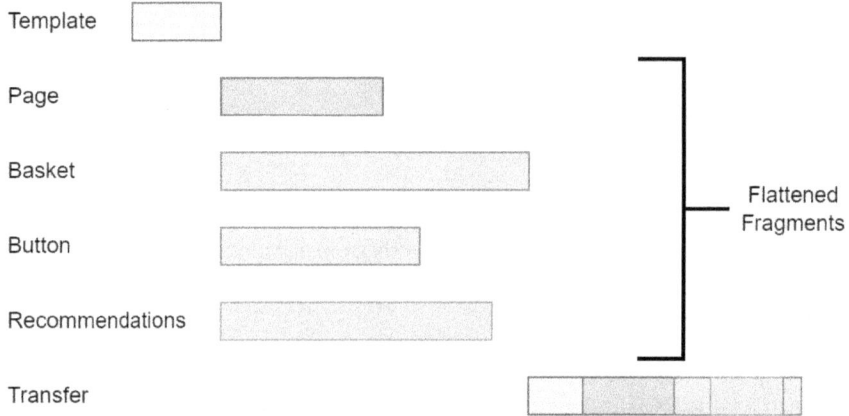

Figure 8.3 – Flattening the fragment's inclusion leads to much better performance

With edge-side composition, we actually drop the template and replace it with the output of the page from the red micro frontend.

In this example, this will flatten the structure. Besides this change, the other thing we must ensure is that the other sources will be cached, if possible. For the basket, we may potentially need a more sophisticated lookup; however, for the other two fragments, the cache could be longer-lived and not need as much logic.

The following diagram contains the final target, as proposed by edge-side composition. It may not reduce the required bandwidth for transfers, but it certainly should trim down response times:

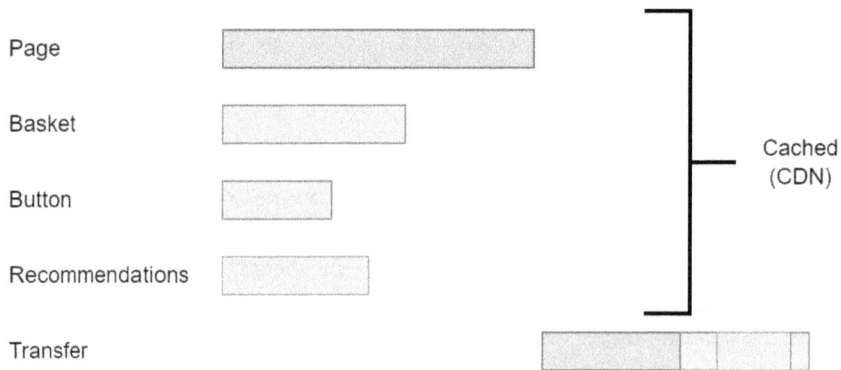

Figure 8.4 – Replacing the template with the main fragment. Resolving
everything from the cache yields the best performance

The major disadvantage of edge-side composition is that it is limited to only a few cases. As we saw in our example, this only worked because we simplified some cases. We also had to post the form directly to the URL of a micro frontend, which is something to avoid in general. Therefore, pretty much all the implementations of edge-side composition are either rather simple or only used on top of other patterns, such as server-side composition.

Edge-side composition is used by a couple of companies. In most cases, these solutions do not exclusively build upon edge-side composition but use it together with some other patterns. One company that leverages edge-side composition heavily is IKEA. Since its website is yet another web shop, the hypothesis is that backend-driven micro frontends make a lot of sense for this kind of problem.

One of the pillars of edge-side composition is its heavy dependency on standard SSI and ESI. We will look at this in more detail in the next section.

SSI and ESI

If you followed the introductory chapter carefully, then you will know that SSI and its successor, ESI, were invented to replace the client-side mechanism framesets with a server-side counterpart. There have been multiple reasons for this:

- Framesets required more requests
- Framesets could not take advantage of the parent's base styling or structure
- Framesets in general make the website inconsistent

Naturally, the first thing that changed was the `<frameset>` element. Since the first impression was to keep this addition hidden on the client in case the backend did not process it properly, another mechanism had to be chosen.

A special comment was born:

```
<!--#include virtual="../quote.txt" -->
```

Some web servers require a special kind of file extension for static files, while others require a configuration for SSI to be active.

SSI

The common SSI directives that are supported by pretty much all web servers – that is, at least Apache, LiteSpeed, nginx, and IIS – are as follows:

- `include` includes the content of a file specified in the `virtual` or `file` attribute
- `exec` places the output of running a program or script located at the `cmd` or `cgi` attribute
- `echo` displays an environment variable defined in the `var` attribute

- `config` configures the display format for the date, time, or file size given by the respective attribute (for example, `timefmt` for time)

- `flastmod` displays the date when a file specified by the `virtual` or `file` attribute was last modified

- `fsize` displays the size of a file specified by the `virtual` or `file` attribute

These are accompanied by control directives that are only supported by a much smaller number of web servers. One of the servers that supports them all is Apache. Here, we have the following:

- `if` evaluates a condition specified in the `expr` attribute

- `elif` evaluates another condition specified in the `expr` attribute when a former `if` directive failed

- `else` takes an alternative content fragment in case of a failed `if` directive

- `endif` must be specified to stop conditional content

- `set` can be used to give a variable named in the `var` attribute a new value specified in the `value` attribute

- `printenv` prints all available variables

While the support on web servers is great, the support on CDN providers is mediocre at best. Here, the more modern alternative, ESI, is more common.

ESI

Many CDN providers support the use of ESI. Among them, the popular choices are Cloudflare, Akamaii, or Fastly. One of the reasons for its widespread support is that ESI is already available in Varnish, which is often part of the infrastructure of these CDNs.

Activating ESI in Varnish can be done using the following code snippet:

```
sub vcl_fetch {
    if (req.url ~ "/*\.html") {
        set beresp.do_esi = true; /* Do ESI processing */
    }
}
```

This will activate edge-side includes in any static HTML page.

Besides being a bit more modern, one of the core reasons for the increased adoption of ESI is that it is based on an official W3C specification, which can be found at `https://www.w3.org/TR/esi-lang/`.

While SSI places all its functionality in a special comment node, ESI uses a standard element with the esi prefix, such as `esi:include`. One advantage of using these tags is to be able to make use of its tree features. For instance, the `esi:try` tag is not self-closed and contains the necessary structure while using `esi:attempt` and `esi:except` inside it as children:

```
<esi:try>
  Ignored markup here...
  <esi:attempt>
    <esi:include ... />
    This line is valid and will be processed.
  </esi:attempt>
  Ignored markup here...
  <esi:except>
    This HTML line is valid and will be processed.
  </esi:except>
  Ignored markup here...
</esi:try>
```

Based on this structural advantage, ESI can cover a lot more cases than SSI. One example that illustrates the power of ESI is conditional elements. Using `esi:choose`, different cases can be iterated with children written as `esi:when`. The default case is denoted by `esi:otherwise`. Now, this seems very similar to `if` conditions in SSI, but the expressions that are used in the test attribute of `esi:when` are a lot more flexible.

Let's look at an example from the official specification:

```
<esi:choose>
  <esi:when test="$(HTTP_COOKIE{group})=='Advanced'">
    <esi:include src=
      "http://www.example.com/advanced.html"/>
  </esi:when>
  <esi:when test="$(HTTP_COOKIE{group})=='Basic User'">
    <esi:include src="http://www.example.com/basic.html"/>
  </esi:when>
  <esi:otherwise>
    <esi:include src=
      "http://www.example.com/new_user.html"/>
  </esi:otherwise>
</esi:choose>
```

The expressions we've used here can use ESI variables such as QUERY_STRING, HTTP_COOKIE, HTTP_HOST, and HTTP_USER_AGENT. Combined, this results in a wide range of possibilities – especially when they're used as variables directly in HTML.

For instance, by entering an `esi:vars` block, we can just write the following HTML code:

```
<esi:vars>
  <img src="/$(HTTP_COOKIE{type})/hello.gif">
</esi:vars>
```

Replacing these special constructs must be enabled via `esi:vars`. Otherwise, the final output of the previous example would literally contain `/$(HTTP_COOKIE{type})/hello.gif` as the image's source.

Finally, if we fear that our ESI decorations may interfere with the standard rendering in case ESI is not available, we can fall back to a special comment section. By wrapping parts of the HTML in `<!—esi ... -->`, we ensure that these parts are only visible when they're being processed by an ESI-compatible web server. In this case, the special comment wrapper will be removed, leaving only ... as the content.

Now, our aggregation layer can finally go to work and compose one website out of multiple sources.

Another way to look at the aggregation layer of an edge-side composition is to combine it with the BFF concept we've heard already in *Chapter 7*.

Stitching in BFFs

The notion of a dedicated aggregation layer is a crucial part of server-side composition and edge-side composition. But going beyond these two patterns, we can see BFFs being used for all kinds of things – not only as aggregation layers to render HTML, but also to provide information that is only relevant for the frontend.

While client-side composition and, in general, stitching in the browser is not uncommon, the potential performance improvements that come by providing everything – potentially even cached – from a central source should not be underestimated.

Bringing in a CDN to serve static resources faster is definitely a good way to gain performance. A good combination of server-side composition and edge-side composition would make use of layouts and advanced HTML manipulation (for example, for forms) on the server and bring together the final pieces in a flat stitching approach.

Flat stitching refers to ESI or SSI resolutions that have a maximum depth of 1. This means that resolving a fragment via ESI or SSI will not result in another HTML piece requiring some resolution.

Summary

In this chapter, you learned how edge-side composition can help us bring micro frontends closer to the client. Using edge-side composition, we get a simpler yet faster alternative to server-side composition. However, besides the raw performance gains, we need to be careful in our micro frontends so that we support the pattern. It is no coincidence that many tools and frameworks that work for server-side composition can be leveraged – at least partially – for edge-side composition, too.

This pattern can be used for information-driven websites, too, under the constraint that the micro frontends are less complex than their server-side composed counterparts.

In the next chapter, we will go one step further by fully composing micro frontends on the client. We'll see that new challenges and opportunities await us.

9
Client-Side Composition

In the previous chapters, you've seen that micro frontends can be very well implemented on the backend, that is, being rendered by individual servers and then either composed via frames or by some aggregation layer. Now, we'll shift a bit and look at how micro frontends can be stitched together on the client. Usually, this means in the browser.

As the example of the tractor store already indicated, server-side composition may bring some challenges for interactive pages. If you recall the issues with the form handling, we could have had a much easier time solving this purely on the client.

In this chapter, we will cover the pattern of client-side composition, which gives us yet another possibility to set up and implement micro frontends.

In general, client-side composition can be introduced in many ways. For this chapter, we'll go with web components. This way, we'll be independent of any specific JavaScript framework, ultimately making the evaluation of the advantages and disadvantages of this pattern a lot more straightforward. If you are not familiar with web components yet, you'll find all the relevant information to use them in this chapter.

Overall, we'll touch on the following topics in this chapter:

- Basics of client-side composition
- Advantages and disadvantages of client-side composition
- Diving into web components
- Composing micro frontends dynamically

By the end of this chapter, you will be able to understand when client-side composition makes sense and how we can implement it practically.

Technical requirements

To follow the code samples in this book, you need knowledge of JavaScript and how to use the command line. You should have Node.js (version 20 or higher) installed using the instructions at `https://nodejs.org`. For this chapter, additional knowledge in using the DOM API can be helpful.

The code used in this chapter can be found in the following repo:

`https://github.com/PacktPublishing/The-Art-of-Micro-Frontends-Second-Edition/tree/main/Chapter09`

The CiA video for this chapter can be found at `https://packt.link/VAupR`

Basics of client-side composition

The beauty of client-side composition lies in its simplicity and directness. There is a lot of modern web philosophy employed here without requiring any complicated backend infrastructure. The main ingredient is usually a framework-agnostic transportation mechanism using web components even though other mechanisms might be used, too.

Framework-agnostic approaches have the advantage that they only use browser APIs, but no custom APIs or principles brought in by some framework. As such, they do not require any maintenance and work well under pretty much all circumstances. They are also among the most secure approaches as browsers actively test and update to have their APIs used the intended way.

We will start by looking at the architecture of client-side composition before we follow the essential steps to come up with a sample implementation. Finally, we will look at potential enhancements for the sample.

The architecture

In the previous chapters, we always relied on a backend component to return a fully composed HTML document to the browser. What if we don't need to compose the document in the backend? What if the browser can take the job of the aggregation layer? As you guessed correctly, the answer lies in the architecture of the client-side composition.

As shown in *Figure 9.1*, client-side composition does not require an aggregation layer. Instead, it only needs to know the location of its primary HTML document containing a script reference. This primary HTML document is often called the application shell or just the **app shell**. The app shell holds references to the most important resources such as the scripts representing the individual micro frontends:

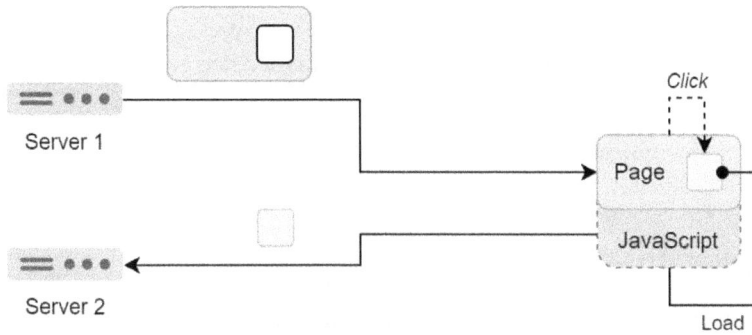

Figure 9.1 – Solutions using client-side compositions rely on JavaScript
to load and present the remaining fragments

Ideally, these scripts are as self-contained as possible. They could be served from dedicated servers or a central content provider. One simple way to publish them is to release them as npm packages, which are then consumed from **Content Delivery Network (CDN)** mirrors such as UNPKG or jsDelivr.

> **Important note**
>
> npm packages are just compressed tarballs. npm packages contain the `package.json` file for metadata, as well as all the non-ignored files from the original directory. An npm package is created with the `npm pack` command. Publishing a package can be done with `npm publish`. Importantly, to publish a package you need to be authenticated using the `npm adduser` command.

While the actual contents of the script can be rather arbitrary, client-side composition usually relies on an established mechanism and fixed pattern. One of the most convenient ways is to use web components.

Using web components gives us a lot of flexibility and freedom. It allows us to introduce a specific micro frontend in any application anywhere. Likewise, web components come with their own problems, which need to be tackled and tamed. We'll look into the specifics of this later in this chapter.

From an architecture point of view, we can choose between these two extremes:

- Taking one script per web component
- Taking one script per micro frontend

While the former is generally a great idea, the latter can be much simpler and provide slightly better performance. Taking one script per web component could result in a lot of complexity to eventually orchestrate things that belong to the same domain.

On the other hand, only having a single script per micro frontend may result in overhead for applications that only need to use a single web component. In the end, this decision goes back to the principles of domain decomposition as outlined in *Chapter 4*.

Let's take a look at a sample implementation – transforming the tractor store to be composed in the browser.

Sample implementation

We take the previous code from *Chapter 7* as a starting point to come up with a suitable implementation of the client-side composition pattern.

For each micro frontend, we will produce one script for each fragment. While the blue micro frontend will end up with two scripts, the other micro frontends will produce only a single script. This was done to have one script per exported component.

Producing these scripts can be done in multiple ways. We could just write them directly, but then we may have issues with references to assets such as images or stylesheets. We can also use tooling such as a bundler.

One of the most popular bundlers is webpack. webpack takes some JavaScript files as entry points and follows their imports to come up with some output files that represent the inputs we've seen in an optimized way. For instance, it will minify and combine JavaScript files into a single asset. Furthermore, it is possible to lazy load additional resources or write out a source map for improved debugging.

In our case, the configuration file for webpack (webpack.config.js) looks as follows:

```
module.exports = {
  entry: {
    'product-page': './src/product-page.js',
  },
  output: {
    filename: '[name].js',
    path: __dirname + '/dist',
    publicPath: 'http://localhost:2001/',
  },
  devtool: 'source-map',
  module: {
    rules: [
      {
        test: /\.(png|svg|jpg|gif)$/i,
        use: ['file-loader'],
      },
      {
        test: /\.css$/i,
```

```
        use: [{
          loader: 'style-loader',
          options: { injectType: 'linkTag' },
        }, {
          loader: "file-loader",
        }],
      },
    ],
  },
};
```

This is pretty much a standard webpack file that is capable of making file references using `file-loader`. `style-loader` can bring in `<link>` tags automatically using a reference to imported CSS files.

The imports to CSS files can be written using the ES module specification:

```
import "./style/basket-info.css";
```

Individual web components such as the `BasketInfo` are just HTML elements that listen for changes to the `sku` attribute.

When we change the `sku` attribute's value, we may need to re-render the web component or change part of its content. One of the easiest – yet also one of the most destructive and least performing – ways is to just set the `innerHTML` property of the custom element. This will remove all child nodes from the element, parse the given HTML content as a fragment, and append the nodes from the resulting fragment to the element.

For `BasketInfo`, the implementation could look as follows:

```
class BasketInfo extends HTMLElement {
  constructor() {
    super();
    this.render();
  }

  static get observedAttributes() {
    return ['sku'];
  }

  render() {
    const count = items.length;

    this.innerHTML = `
<div class="${count === 0 ? 'empty' : 'filled'}">
```

```
      basket: ${count} item(s)</div>
      `;
  }

  attributeChangedCallback(name, oldValue, newValue) {
    if (name === 'sku' && oldValue !== newValue) {
      this.render();
    }
  }
}
```

Before we can really use a custom element, we also need to give it a name. By convention, the naming should follow the kebab-casing style. In kebab-case, identifiers are all in lowercase with minus symbols connecting the different segments. For the `BasketInfo` class, this results in the following:

```
customElements.define('basket-info', BasketInfo);
```

Especially for the interplay of the `basket-info` and `buy-button` components, we need to have a common mechanism. Luckily, both components come from the same micro frontend, but since they are independently deployed and mounted, we still need a way of communicating state changes here.

Earlier, we used a session together with a page refresh to update the basket accordingly. Now we can just use DOM events to communicate a change. When the button is pressed, we fire a custom event called `add-item`. The basket will listen to this event and change its state. To notify potentially interested UI fragments of this new state change, another custom event, `added-item`, is dispatched.

In code, the state logic in the basket module can be written as follows:

```
const items = [];

window.addEventListener('add-item', () => {
  items.push("...");
  window.dispatchEvent(new CustomEvent('added-item', {
    detail: items,
  }));
});
```

Finally, we need to define the app shell for our client-side composition. Since we know that there is a `product-page` in the application, we will go ahead and place that in the body. Other than that, we need to include the scripts for all components.

The scripts are deployed on individual web servers, which could be managed by the individual teams. In the example implementation, we just refer to their port on `localhost`:

```
<body>
  <product-page id="app"></product-page>
  <script src="http://localhost:2001/product-
    page.js"></script>
  <script src="http://localhost:2002/basket-
    info.js"></script>
  <script src="http://localhost:2002/buy-
    button.js"></script>
  <script src="http://localhost:2003/product-
    recommendations.js"></script>
</body>
```

Rendering this in the browser results in a DOM as shown in *Figure 9.2*. We see that our previous `<div>` wrappers for the individual fragments have been replaced with `custom` elements and the overall structure is still the same:

```
▼ <body>
    ▼ <product-page id="app" sku="eicher"> grid custom…
        <h1 id="store">The Model Store</h1>
      ▶ <basket-info id="basket" class="blue-basket" sku="eicher"> ⋯ </basket-info> custom…
      ▶ <div id="image"> ⋯ </div>
      ▶ <h2 id="name"> ⋯ </h2>
      ▶ <div id="options"> ⋯ </div> flex
      ▶ <buy-button id="buy" class="blue-buy" sku="eicher"> ⋯ </buy-button> custom…
      ▶ <product-recommendations id="reco" class="green-recos" sku="eicher"> ⋯ </product-
        recommendations> custom…
    </product-page>
    <script src="http://localhost:2001/product-page.js"></script>
    <script src="http://localhost:2002/basket-info.js"></script>
    <script src="http://localhost:2002/buy-button.js"></script>
    <script src="http://localhost:2003/product-recommendations.js"></script>
</body>
```

Figure 9.2 – The DOM generated by the app shell

The app shell is therefore fully composed of these elements. In *Figure 9.3*, we see that the composition worked quite naturally, starting with the `product-page` component that relied on components such as `basket-info`, `buy-button`, and `product-recommendations`.

The components had to be brought in by their scripts first:

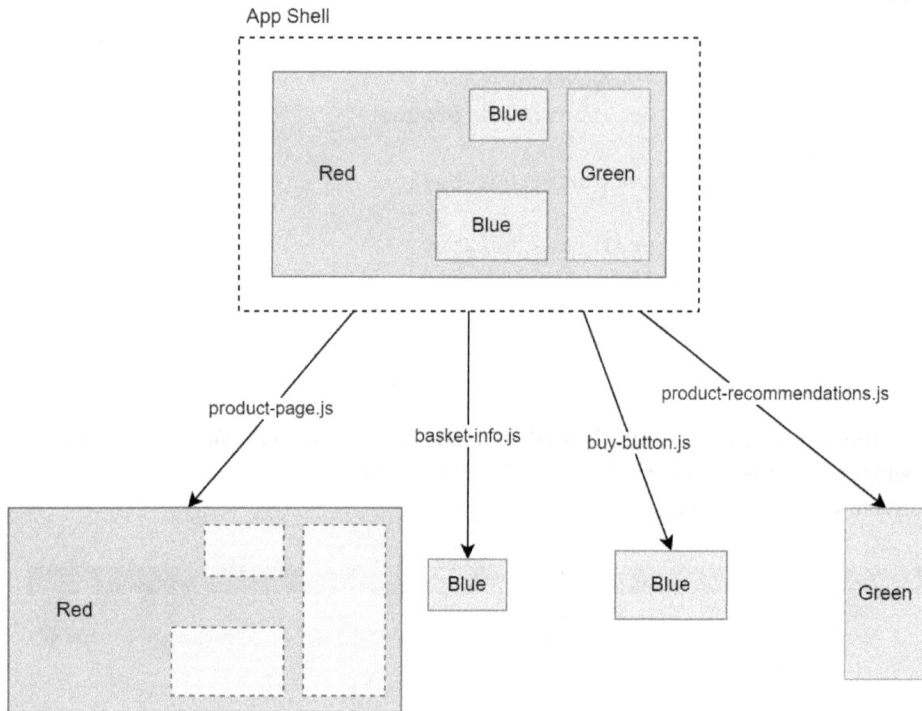

Figure 9.3 – The app shell composes the frontend from individual web components

Since websites will never fail directly, web components by themselves also don't crash. For instance, when the server providing `product-recommendations` goes down, all that's happening is that `product-recommendations` will remain an empty placeholder.

On the other hand, it's pretty difficult to detect this case making the use of fallbacks more difficult than it should.

> **Important note**
>
> Client-side composition is quite often used as an enhancement to server-side rendered pages. In the case of single-page applications, where the full rendering is done on the client, a routing engine has to be used, too. While framework-specific solutions exist, a framework-independent approach should be picked. One solution is to use `universal-router`. More information is available at `https://kriasoft.com/universal-router/`.

With the app shell being so simple, there must be many potential enhancements, right? Let's have a look.

Potential enhancements

Looking at potential enhancements, we directly see a couple of things that may seem odd. For instance, the app shell needs to include all script references. Ideally, the app shell would only need to know the location of a web component directory, which would be used to automatically load web components that are actively used. This way, micro frontends could use whatever web components they like, with the app shell loading the respective scripts when they are needed.

Another thing to consider is to boost the responsibilities of the app shell. In this simple example, we only had a basic HTML structure, its core styling, and the script references in the app shell. Things such as authentication, logging, or error handling are not included, but could be viable additions.

Finally, we should not require a single server per micro frontend. With these static files in mind, most micro frontends will only need a way to serve their static files. In this case, a simple CDN-like mechanism is more than sufficient. As a beautiful side effect, we would require fewer DNS lookups and be able to leverage modern transport mechanisms.

What advantages and disadvantages does this approach give us? Let's find out.

Advantages and disadvantages of client-side composition

The most obvious advantage – and disadvantage – is that client-side composition relies on JavaScript. This leads to performance challenges and accessibility issues. There is nothing here that cannot be improved, but rather things that need to be considered wisely and taken care of.

The web component standard itself is a widely implemented standard that focuses on the basics rather than fancy abstractions. Clearly, this means that changes are likely to never break existing implementations. However, it also means that other frameworks will be placed on top of it and that these frameworks are most likely a more productive basis for development than using web components directly.

If we want to really leverage one of the key features of web components, namely **shadow DOM**, we cannot support outdated legacy browsers such as Internet Explorer. By itself that would not be such a big deal, however, since web components provide no standardized way of being rendered on the server, we may have a deal-breaker here.

Client-side composition is a pattern that can be found quite often, just like some other patterns from the previous chapters, as an addition to an existing architecture. In fact, using it with web components makes us feel that the adoption of web components goes somewhat hand in hand with the rise of micro frontends.

Usually, client-side composition is a pattern that can be mostly found in tool-like web applications. These are web apps that – in contrast to websites mostly focused on providing information – are highly interactive.

Even though web components are not the only way to achieve client-side composition, they are by far the most popular technique. Let's use this opportunity to strengthen our knowledge about web components.

Diving into web components

With the rise of JavaScript frameworks and the increased importance of client-side rendering, web components have been proposed and standardized. The term *web component* is used as an umbrella expression for a set of features that try to provide a standard component model for the web.

In the following sections, we'll try to understand the most important features that actually form web components and how web components can help us isolate styles.

Understanding web components

As you know already, web components provide the ability to use custom elements in HTML documents. These could look as simple as the following snippet:

```
<product-page id="app"></product-page>
```

While custom elements cannot be changed from the HTML parsing perspective, they can be fully configured in terms of behavior and appearance – for instance, custom elements cannot be self-closed, such as `<img>` or `<meta>` tags. To add some custom appearance, shadow DOM can be used.

Besides custom elements, the term *web component* refers to many other features. The most important features are the following:

- **Shadow DOM** to encapsulate DOM trees
- **HTML templates** to provide reusable DOM trees
- **Custom elements** to define new elements

In the previous sample, we exclusively used custom elements even though HTML templates could have been used, too. The main advantage of using HTML templates is that they provide the DOM a native way of actually copying and reusing HTML fragments. Without HTML templates, fragments could be used, too, but only programmatically from JavaScript – not declaratively within the document's HTML source.

> **Important note**
>
> The browser support for web components is fairly good. The custom element specification is supported in all major browsers except the no-longer-supported Internet Explorer. If unsupported browsers really need to be handled, there is usually a solution in form of a small library to enable the otherwise missing functionality.

Custom elements expose the life cycle of HTML elements. They allow hooking into mounted and unmounted events, as well as being notified when an attribute changes. Without these life cycle events, we'd have to fall back to using a `MutationObserver`.

Quite often web components are not used directly, but rather indirectly as a transport mechanism for another application. We've seen that the `product-page` element of our sample could be considered as its own mini-application. Going forward, we may have used another framework such as React inside the web component. For this, we can use the life cycle methods of web components as illustrated in *Figure 9.4*:

Web Component

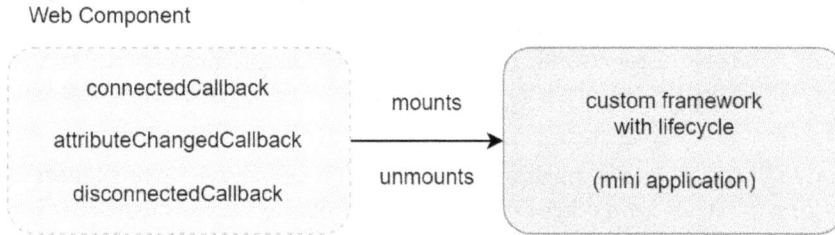

Figure 9.4 – Mini applications using arbitrary frameworks can be managed by web components

When using `attributeChangedCallback`, we need to make sure to communicate the attributes to watch via an `observedAttributes` property on the custom element's class.

Last, but not the least, shadow DOM is one of the most interesting things about web components. This is the magic ingredient that makes independent development of micro frontends using web components possible in the first place.

Isolating styles with shadow DOM

One of the advantages that web components bring is direct access to shadow DOM. Shadow DOM has two modes, `open` and `closed`:

- `open` indicates that the tree below the `shadowRoot` is open for modification and can be accessed on the `shadowRoot` reference of the element
- `closed` indicates that the tree below the `shadowRoot` is closed for modification and cannot be accessed unless the reference was stored by its creator

Quite often, the additional encapsulation gained by using `closed` is not that interesting. In open mode, we already have access to a very nice feature of shadow DOM: style isolation.

Style isolation is crucial to prevent styling conflicts between micro frontends. Without shadow DOM, we'd require techniques such as CSS namespacing to guard the layout on one micro frontend from being impacted by the styles from another micro frontend. Furthermore, it prevents global styles from leaking, which improves reliability, too.

On the downside, shadow DOM has quite weak support in older browsers and cannot really be polyfilled. To use shadow DOM, JavaScript is required, which is not a problem for client-side composition alone, but potentially not an option in combination with some server-side rendering. Here, we would need support for declarative shadow DOM, which is a rather recent specification to allow SSR of a shadow DOM definition.

> **Important note**
>
> Polyfills play a central role in bringing modern web features to older browsers. Quite often, support for browsers such as Internet Explorer 11 is required even though these browsers have limited support for many features that are demanded in modern web applications. A polyfill is a piece of code that enables browsers without the support of a specific feature to provide it anyway. While many features can be polyfilled, some are just too low-level to provide a suitable fallback.

Style isolation itself comes with some drawbacks, too. With global styles being ignored, we are not only ignoring the styles from other micro frontends, but also the ones that actually give the application a consistent look and feel. Reusing some common styling is quite cumbersome and requires explicit re-imports.

Going beyond the strengths and drawbacks of shadow DOM, we still have to discuss how to actually use micro frontends dynamically; that is, without having a predefined list of scripts in the app shell. We will discuss this in the next section.

Composing micro frontends dynamically

We briefly introduced the idea of using a central web component registry to lazily load the different scripts required to make the web components work. This would indeed be the ultimate solution, as it is lazy yet dynamic. However, to make this fully operational is also not easy. An easier way is to start with a micro frontend discovery service.

Using a micro frontend discovery service

A micro frontend discovery service is similar to the link dictionary introduced in *Chapter 6*. However, instead of a server resolution, it would just return paths to all the available scripts.

For example, the following API response could be triggered for the tractor store sample:

```
{
  "scripts": [
    "http://localhost:2001/product-page.js",
    "http://localhost:2002/basket-info.js",
    "http://localhost:2002/buy-button.js",
    "http://localhost:2003/product-recommendations.js"
  ]
}
```

The great thing about this response is that it can be adjusted for a respective user or use case. If the app shell is delivered for an anonymous user, a different script delivering a product-page component could be provided.

Once we have such a dynamic backend response, we can integrate it in the frontend. This can be implemented in multiple ways:

- Already fully assembled in the index.html file with each script being placed there as a valid <script> tag

- Fetched and loaded dynamically from a small loader JavaScript embedded in the index.html file

The loader JavaScript may be as simple as the following code:

```
fetch('http://localhost:1234/microfrontends')
  .then(res => res.json())
  .then(({ scripts }) => scripts.forEach(url => {
    const script = document.createElement('script');
    script.onload = () => {}; // success
    script.onerror = () => {}; // error
    script.src = url;
    document.body.appendChild(script);
  }));
```

While the whole construct is capable of providing the micro frontends – or more specifically, their web components – dynamically, it does not prevent using wrongly named custom elements or failing to use the interface of the custom elements correctly.

Quite often, problems with micro frontends could be patched easily at runtime. This is quite simple for the runtime of the web server, but even possible for the runtime in the client. Let's see how.

Updating micro frontends at runtime

One of the promises of micro frontends is that they simplify update paths. Instead of needing to restart a full web server or application, we can just restart a single module – either represented by one of many servers or a module of frontend assets.

Looking at client-side composition, the runtime aspect is fully given by a single user having a browser running the application. Updating one micro frontend should therefore be operational – at least theoretically – on a part of the application without having to forcefully refresh the page by pressing *F5*. This type of behavior is called a hot reload as a part of the website is reloaded after it was already fully loaded. In contrast, a full reload could be referred to as a cold reload.

Unfortunately, web components are not the best candidate for runtime patching. The major problem here is that only a single definition of a web component can be done – with no way to undo that. So why have this discussion then, if it's not possible?

As it turns out, there are ways to actually hot reload a web component. For instance, by keeping the original reference, we could just patch the life cycle methods of the existing class. This, however, has some drawbacks:

- Existing instances don't re-run constructors

- Existing instances will not get added fields and properties

- The `observedAttributes` list cannot be updated

A better way would be to actually place an abstraction on top of the `define` function of `customElements`. The abstraction would actually work, as the following pseudocode illustrates:

```
const elementsVersion = {};

function defineMfElement(elementName, Component) {
  const currentVersion = elementVersions[elementName] || 0;
  const version = currentVersion + 1;
  elementVersions[elementName] = version;
  customElements.define(`${elementName}-${version}`,
    Component);

  if (version === 1) {
    class WrapperComponent extends HTMLElement {
      connectedCallback() {
        // attach event handler to rerender in case of
        // update
        // render the element; forward all attributes
      }
      disconnectedCallback() {
        // remove event handler
      }
    }

    customElements.define(elementName, WrapperComponent);
  } else {
    window.dispatchEvent(new CustomEvent('component-
      updated', { detail: elementName }));
  }
}
```

The only challenge left is that an attribute-changed callback requires an explicit list of attributes to watch. We can solve this by introducing a flat mutation observer in the wrapper, which will properly forward all the attribute changes to the wrapped element.

With these changes, we can dynamically update web components – making a true hot reloading of micro frontends possible. Generally, however, we should be a bit defensive here, as there are some UX challenges on this one, too. We will discuss this in more depth in *Chapter 18*.

Summary

In this chapter, we learned how micro frontends can be composed together in the browser. We saw how web components can be used to transport pieces from different sources statically and dynamically together. We also discussed the advantages and challenges that come with this pattern.

The client-side composition makes the most sense for web applications that come with a few widgets that run independently from the content. One example here would be an online newspaper that publishes articles where further info can be found in interactive graphics. It is best used to extend existing applications or to migrate them to micro frontends slowly.

The information presented in this chapter should help you decide in favor or against using the client-side composition pattern. The code that we've used for doing a sample implementation can be used as a starting point for creating a micro frontend solution.

In the next chapter, we will look at a refinement of this pattern, which allows single-page applications to leverage micro frontends. We'll see that some of the shortcomings of web components can be eradicated by using more sophisticated frameworks instead.

10
SPA Composition

In the previous chapters, we had a glimpse of how micro frontends can be composed just using the browser without any active backend support. Most of the time, this technique becomes really interesting with increased user interactivity.

Today, **single-page applications** (**SPAs**) offer a rich interactive application feeling that many experiences target. Combining multiple components from standard SPA frameworks such as Angular or React is described with the SPA composition pattern.

The key factor for a SPA is that routing is not done on the server, but rather in the frontend. Routing in JavaScript requires techniques such as taking the URL's hash or controlling the history API of HTML5. Then older features such as AJAX can be used to lazy load the content that should be presented on a page.

In this chapter, we cover the pattern of SPA composition, which allows different SPA frameworks to be contained in the same application.

We'll touch on the following topics:

- Basics of SPA composition
- Advantages and disadvantages of SPA composition
- Building a core SPA shell
- Integrating SPA micro frontends
- Exploring communication patterns
- Optimizing hydration and progressive rendering

This should help us to understand when SPA composition makes sense and how we can implement it practically.

Technical requirements

To follow the code samples in this book, you need knowledge of JavaScript and how to use the command line. You should have Node.js (version 20 or higher) installed using the instructions at `https://nodejs.org`. For this chapter, additional knowledge of using the webpack bundler can be helpful.

The code used in this chapter can be found in the following repo: `https://github.com/PacktPublishing/The-Art-of-Micro-Frontends-Second-Edition/tree/main/Chapter10`

The CiA video for this chapter can be found at `https://packt.link/URmhU`

Basics of SPA composition

SPA composition builds on top of client-side composition. However, instead of explicitly mentioning HTML tags, we just load the micro frontends in form of some script files. These scripts will automatically integrate the contained components, mounting and rendering them when certain conditions are fulfilled. In case of fulfilled conditions, we refer to those sub-applications (independent or little SPAs) as being active. The trick is that these independent SPAs will not always be active – and if they are active, they may coexist within one another.

As with all the other patterns, we start our journey into SPA composition with a look at its architecture. Then we follow the essential steps to come up with a sample implementation. Finally, we look at potential enhancements for the derived sample.

The architecture

We've already seen how mini applications can be hosted in web components. SPA composition takes this idea a step further. Instead of relying on web components to handle the lifecycle of mini applications, a custom loader script is used.

As shown in *Figure 10.1*, the pattern is quite similar to client-side composition. However, instead of delivering an HTML page that comes with holes for fragments, SPA composition is deployed with an app shell that includes the previously mentioned loader script.

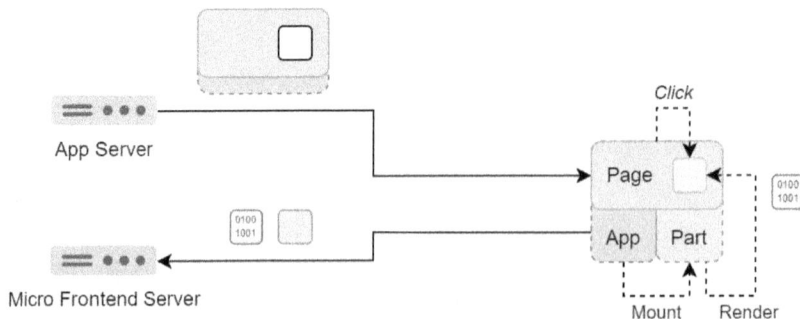

Figure 10.1 – SPA composition uses a loader script to orchestrate different SPA frameworks

The job of the loader script is to orchestrate the different micro frontends. Each micro frontend represents an independent SPA. However, instead of always being active and capturing all the navigation, each micro frontend only turns on its SPA when certain conditions are met. These conditions can be anything, from a specific URL to a certain state, to an anticipated user action.

To justify SPA composition, a more dynamic and interactive application than the tractor store should be used. We'll create a small accounting app consisting of multiple pages distributed over three micro frontends. Let's look at the details.

Sample implementation

For this sample, we'll create an accounting application that comes with three micro frontends. From a domain decomposition point of view, we get the following modules:

- Balance, concerned with displaying the balance sheet, adding income or outgoings entries, and viewing details
- Tax, concerned with indicating items that are tax-deductible
- Settings, concerned with the user's preferences

All cross-cutting concerns such as currency handling, the application's layout, and the menu structure are placed in the SPA shell. In total, the SPA comes with three dedicated pages:

- The balance sheet (from the balance micro frontend)
- Details of an entry (from the balance micro frontend)
- The current settings (from the settings micro frontend)

For the design system, Bootstrap v5 has been chosen. The beauty of this choice is that Bootstrap is by default framework-agnostic. As such, all micro frontends can use Bootstrap if the SPA shell integrates it.

For the shell, we don't require much. We choose a small HTML document, which contains references to the application's stylesheet and script elements. Also, the basic layout is defined here. Within this layout, we'll find a container for content coming from the micro frontends:

```
<main class="col-md-9 ms-sm-auto col-lg-10 px-md-4"
    id="app-content"></main>
```

In SPA-like fashion, we have the full application defined inside the application's script. From a high-level perspective, it looks as follows:

```
// Registers a new component with the lifecycle functions
window.registerComponent = (appName, componentName,
    lifecycle) => {};
```

```
// Renders the component in the target element
window.renderComponent = (appName, componentName, target,
  props) => {};

// Destroys the component in the target element
window.destroyComponent = (appName, componentName, target)
  => {};

// Activates the component when the URL changes
window.activateOnUrlChange = (appName, componentName,
  handler) => {};

window.addEventListener('popstate', urlChanged);

import('./scripts.json').then(scripts =>
  scripts.default.forEach(url => {
    const script = document.createElement('script');
    script.src = url;
    document.body.appendChild(script);
}));
```

The global functions can be used in the micro frontends to bring in additional content. For instance, the registerComponent function can be used to declare new components. These components can then be rendered using the renderComponent function. The activateOnUrlChange function is used for inserting a listener in the routing engine.

The global functions are all part of the previously mentioned loader script, which orchestrates the different micro frontends. To do this, the routing engine in the loader script also needs to couple to DOM events such as popstate for being informed when the URL of the browser changes by pressing the back button.

To make the routing engine convenient and truly universal, we attach a listener following all SPA internal links:

```
document.body.addEventListener('click', followLink);
```

This way, most micro frontends will not even need to know about the routing engine and its implementation. They just use links for navigation, which will then remain internal in the application instead of resulting in a full page reload.

The final part of the code snippet deals with loading the different micro frontends. In SPA composition, each micro frontend is primarily represented by a valid JavaScript file. We collect – and load – all micro frontends in a JSON file, `scripts.json`, which can look as simple as this:

```
[
  "https://example.com/mfs/balance/1.0.0/root.js",
  "https://example.com/mfs/settings/1.0.0/main.js",
  "https://example.com/mfs/tax /1.0.0/index.js"
]
```

Theoretically, this file could be written and updated with some tooling. For this example, we'll maintain it manually to avoid having the actual process hidden by some tool.

Going over to the different micro frontends, these could all be written using different frameworks. For instance, we could have the following:

- The team creating the balance micro frontend is most experienced in React
- The team creating the tax micro frontend likes to experiment with Svelte
- The team creating the settings micro frontend is a huge proponent of Vue

Ideally, most teams use the same framework. This will lead to more consistency and easier maintenance. Nevertheless, for our example, we'll go with the patchwork approach to demonstrate the power and flexibility of this concept.

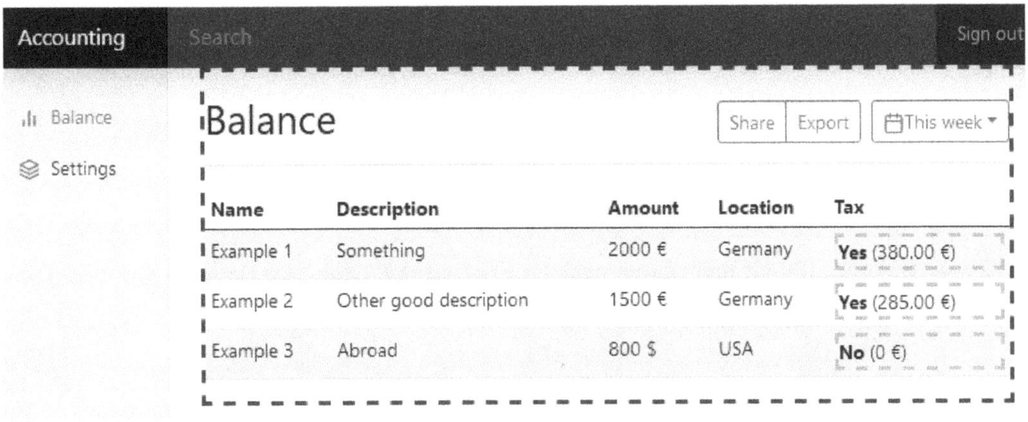

Figure 10.2 – Home page of the accounting example application using SPA composition

The global integration helpers can then be used directly within the micro frontends. For instance, the tax micro frontend only needs to register a single component. It does it like this:

```
let Info = undefined;

window.registerComponent('tax', 'info', {
  bootstrap: () =>
    import('./Info.svelte').then((content) => {
      Info = content.default;
    }),
  mount: (target, props) => new Info({
      target,
      props,
  }),
  unmount: (_, info) => info.$destroy(),
});
```

The module-global variable Info is used to buffer the actual component, which is lazy loaded in the bootstrap function. The mount and unmount functions can then rely fully on the Svelte framework. This offers components as classes with the shown constructor. An instance of a Svelte component comes with a $destroy method to unmount the Svelte component.

The balance micro frontend has been written in React. It uses the tax info component with the lifecycle defined previously. In the end, the code for displaying a row on the balance sheet is as simple as this:

```
const BalanceItem = ({ item }) => (
  <tr>
    <td>{item.name}</td>
    <td>{item.description}</td>
    <td>{item.amount}</td>
    <td>{item.location}</td>
    <td>
      <TaxInfo {...item} />
    </td>
  </tr>
);
```

Here, TaxInfo is a wrapper around calling the global renderComponent and destroyComponent helper functions. The most straightforward implementation for this could look like this:

```
const TaxInfo = (props) => {
  const ref = React.useRef(null);

  React.useEffect(() => {
    // Render the tax info component on mounting
```

```
    window.renderComponent('tax', 'info', ref.current,
        props);

    return () => {
        // Destroy the tax info component when unmounting
        window.destroyComponent('tax', 'info', ref.current);
    };
  }, []);

  return <slot ref={ref} />;
};
```

This approach will only work if the component has been registered already. Deferred registrations will not work. Instead, either events or some other mechanism should be used to ensure an update of the rendering when the component registry changes. Ideally, this is handled already within the framework.

The balance sheet itself is placed on a dedicated page. Therefore, it must be part of the routing engine. Its root module must contain a call to the `activateOnUrlChange` function:

```
import * as React from 'react';
import { render } from 'react-dom';

let BalanceSheet = undefined;

window.registerComponent('balance', 'sheet', {
  bootstrap: () =>
    import('./BalanceSheet').then((content) => {
      BalanceSheet = content.BalanceSheet;
    }),
  mount: (target) => render(<BalanceSheet />, target),
  unmount: (target) => render(null, target),
});

window.activateOnUrlChange(
  'balance',
  'sheet',
  (location) => location.pathname === '/'
);
```

Again, we are using the same technique for buffering the lazy-loaded component. Another mechanism could have been to use React's `lazy` function with a `Suspense` component. In this case, the registered component is only shown when users visit the home page presented at the root path `/`.

Finally, the settings micro frontend works almost identically to the balance one. Besides using a different frontend framework (Vue instead of React), it also does not rely on any component from other micro frontends. Going to the /settings route will unmount the current micro frontend and mount the content coming from the settings micro frontend.

There's plenty to discuss about this pattern, so let's start the discussion with potential enhancements of the sample itself.

Potential enhancements

One of the major issues in this solution is that all micro frontends need to be known by the shell in the form of a JSON file. If a new micro frontend is added, the JSON file needs to be changed, too. Like with all the previous approaches, we could try to decouple that by moving, for instance, the resolution of the micro frontends to a dedicated backend service.

There is another advantage of decoupling the micro frontends from the app shell. Let's say only a few of the micro frontends may fit the current user – why should we still reference or prepare all of them? Instead, why don't we just bring in those micro frontends that have a valid chance of being used? This is also one of the crucial differences of micro frontend solutions that are primarily composed in the backend – we don't need to work with all micro frontends all the time.

In the context of more flexibility, the menu items – or at least all kinds of UI fragments that are related to the functionality scattered across all micro frontends – should not be hardcoded in a SPA shell. Instead, a mechanism such as extending components should be used. We'll discuss this approach later in this chapter.

These enhancements for our sample implementation aside, let's find out what advantages and disadvantages usually come with this pattern.

Advantages and disadvantages of SPA composition

Micro frontend solutions based on the SPA composition pattern share many of the advantages and disadvantages of non-micro frontend SPAs. For instance, one advantage of a SPA is that the interaction feels smooth and immersive. However, since SPAs usually require a lot more resources and JavaScript to work, the initial loading time makes it bulky, too.

One crucial difference to monolithic SPAs is that SPA micro frontends are a lot harder to debug. While monolithic SPAs have great debugging tools, a micro frontend SPA is already difficult to develop. Most solutions will enable some kind of development mode on a live instance – sometimes even the one running in the production environment. While for most development efforts this may be acceptable, the lack of an offline-first development environment can be quite painful at times, e.g., in cases where internet connectivity is down or when you are travelling. In the next chapter, we'll see how this can be improved right away.

In comparison to other micro frontend patterns, the SPA composition is one of the most – if not *the* most – difficult frameworks to render on the server. While this may not seem like a big deal (and usually it is not), under some circumstances we might find this to be a real issue. Almost certainly this excludes this pattern from information-heavy pages such as e-commerce websites or newspaper websites.

In the end, the main targets for using the SPA composition pattern are applications or tools that require heavy interactivity, where the individual parts are either taken from existing applications using different frameworks, or it cannot be guaranteed that all parts will be written using the same frontend framework.

With these advantages and disadvantages in mind, let's explore in detail what needs to be considered when building an app shell within the SPA composition.

Building a core SPA shell

Like in client-side composition, we'll require an app shell to bootstrap and compose the application in the user's browser. Quite often, this shell is rather small and only focused on getting essentials like the routing mechanism for activating pages or a mechanism to share dependencies in place.

In this section, we'll look at these two basic concerns that are inherent to most SPA shells. We'll start with the activation mechanism.

Activating pages

One of the core features of a SPA is routing. As such, a SPA shell should be able to determine what pages are shown based on some conditions. In the sample, we've already seen that a route handler could just be based on the standard DOM API circulating around the `history` object.

In *Figure 10.3*, we see the standard pattern that most routing engines will follow. An activator module is globally installed, which can be used to register or unregister micro frontends. On every route change, the different micro frontends are checked.

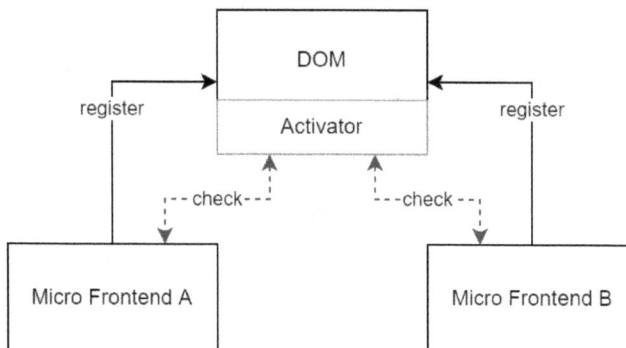

Figure 10.3 – A common activator checks all registered micro frontends for their status

Only in the event of a state change, that is, if an active micro frontend becomes inactive or vice versa, the activator needs to mount or unmount a SPA. Most state changes are just URL changes. And as we know, URL changes are handled by routing engines.

A popular choice for such a routing engine is the `single-spa` package. Essentially, it provides a framework-agnostic router that is coupled to a SPA activator. This allows orchestrating multiple SPA instances in a single application.

> **Important note**
>
> `single-spa` started as a small helper library to help running different SPA frameworks within one application. Later, the project pivoted a bit and repositioned itself as a micro frontend framework. While it comes with many important technical foundations for a micro frontend framework to realize a SPA composition, many parts necessary for a full solution still need to be implemented when choosing `single-spa`.

In `single-spa`, the registration step is done via the `registerApplication` function. Quite often, this is done from a central module when the SPA shell starts. In `single-spa` terminology, such a module is known as a **root config**.

Let's look at an example of a root config in `single-spa`:

```
import { registerApplication, start } from 'single-spa';

registerApplication(
  // name of the app
  'my-app',
  // lazy loading the lifecycle of my-app
  () => import('my-app'),
  // checks if my-app should be active
  (location) => location.pathname.startsWith('/my-app')
);

start(); // starts the application using single-spa
```

The previous code utilizes the pattern described in *Figure 10.3*. First, we call the `registerApplication` function, which provides two important things:

- A function used for lazy loading the actual application (exporting the application's lifecycle)

- A function that can be used for checking whether the particular application should be active at all

The first time the application should be active, the orchestrator needs to bootstrap it. In `single-spa`, this is a combination of triggering the loading of the application, waiting for it to be loaded, and then using the optional `bootstrap` lifecycle export to finalize it. Once all that is done, the `mount` lifecycle is called.

In order to do this job, the orchestrator in the app shell needs to keep track of what micro frontends are actually loaded, still loading, should be mounted, or need to be unmounted. Any change of the URL has to be monitored and leads to a potential state change of the currently active micro frontends.

In this space, another problem arises. Let's say we have three micro frontends – all using the same library, for instance, React. Surely, all three could just bundle React and start React on their own. However, this means three instances of React with required libraries such as React DOM, maybe even React Router, and React Router DOM… That's a lot of libraries! To improve this situation, we may want to share React and only load it once. A typical SPA shell should actually offer a mechanism to allow that. But how? Let's see some strategies.

Sharing dependencies

There are some quite popular mechanisms for sharing dependencies. The easiest mechanism is to declare some dependencies as shared on the app shell already. Using the previous example, we could just declare React and all useful helper libraries as a dependency on the app shell – to be used by the micro frontends.

The challenge with this approach is that each micro frontend needs to know what libraries are shared and how to actually get the shared library instead of a bundled version. Using webpack as a bundler, we can leverage the `externals` field to use a shared library.

In a `webpack.config.js` file, this looks similar to the following:

```
module.exports = {
  externals: ['react', 'react-dom', 'react-router-dom'],
};
```

In the preceding case, webpack will resolve the packages at runtime with a call to the output's module system. This may fail, for example, if there is no `define` function globally available or if the three packages are not registered with the identified module system.

We could be a bit more explicit by using an object such as the following:

```
module.exports = {
  externals: {
    'react': 'React',
    'react-dom': 'ReactDOM',
    'react-router-dom': 'ReactRouterDOM',
  },
};
```

In this case, the libraries are retrieved from global variables such as `React` or `ReactDOM`. Here, the SPA shell would just set `window.React` to contain the exports of the React package.

However, having these global variables is bad – and ideally, we'd just like to stick to the array, or have an implicit way to solve this without requiring explicit configuration. One option to stick to the array is to use the **ES Module (ESM)** system. Unfortunately, support for this feature requires quite recent versions of browsers again.

While Chrome (since version 79), Firefox (since version 67), and Edge (since version 79) provide support, outdated browsers such as Internet Explorer (including version 11) are problematic. There is also no direct polyfill for this. Instead, we could use an alternative module system that resembles ESM quite perfectly but also brings support for pretty much all browsers.

A suitable choice for a working module system is **SystemJS**. SystemJS allows the SPA shell to register shared dependencies in multiple ways.

One option is to directly declare these dependencies within the shell itself:

```
function registerModules(modules) {
  Object.keys(modules).forEach((name) =>
    registerModule(name, modules[name]));
}

function registerModule(name, resolve) {
  System.register(name, [], (_exports) => ({
    execute() {
      const content = resolve();

      if (content instanceof Promise) {
        return content.then((innerContent) =>
          _exports(innerContent));
      } else {
        _exports(content);
      }
    },
  }));
}

registerModules({
  'react': () => require('react'),
  'react-dom': () => require('react-dom'),
  'react-router-dom': () => require('react-router-dom'),
});
```

Another option is to leverage a neat feature called **import maps**. Import maps are an official standard that are directly supported by most browsers using ESMs. Additionally, they can be used with SystemJS, too.

The import map is usually referred to in the shell's HTML code:

```
<script type="systemjs-importmap" src="https://example.com/import-map.
json"></script>
```

The contents of the import map for our example could look like this:

```
{
  "imports": {
    "react": "https://unpkg.com/react/umd/react.
      production.min.js",
    "react-dom": "https://unpkg.com/react-dom/umd/react-
      dom.production.min.js",
    "react-router-dom": "https://unpkg.com/react-router-
      dom/umd/react-router-dom.min.js"
  }
}
```

The great advantage of this approach is that these shared dependencies are not directly bundled into the shell. Instead, they are lazy loaded by SystemJS on demand, which is in effect when a micro frontend is loaded that requires them.

For our micro frontends to work in this case, we need to set the output to use SystemJS, too. In webpack, this is an easy tweak in the configuration:

```
module.exports = {
  output: { libraryTarget: 'system' },
  externals: ['react', 'react-dom', 'react-router-dom'],
};
```

Now, in this case, everything works as expected and is in good shape, except that the import map may already need to know about these shared dependencies. What if a shared dependency is dropped from the import map? What if we don't want to use the version from the import map? What if we always want to align bundled dependencies with the available ones from the import map?

While some of these questions could certainly be addressed within SystemJS, it is also a good option to look for alternatives. The option we can mention in the scope of this chapter is to use **Module Federation**. This is a core plugin for webpack that actually addresses these problems and can be thought of as `externals` on steroids. It allows direct sharing of webpack chunks, as well as adding or removing chunks if applicable. In short, it provides the basis to build an application from federated independent modules.

More details on the topic of dependency sharing can be found in *Chapter 14*.

Now that we can share dependencies efficiently, it's all about integrating these SPA micro frontends.

Integrating SPA micro frontends

As we've seen in the case of `single-spa`, a so-called root config is used to centrally declare the micro frontends – at least their name, activation function, and loader function. In other approaches, we follow a similar concept, usually keeping some information about the different micro frontends in a central location – which is usually in the SPA shell – while being able to use independent deployment pipelines for the micro frontends in general.

This still leaves us with two problems: how do we declare the lifecycle, and how do we now use these cross-framework components? Let's explore both, starting with lifecycle declaration.

Declaring the lifecycle

We've already touched on the lifecycle of the components exported by a SPA micro frontend briefly. In `single-spa`, this boils down to four different functions:

- The `bootstrap` function is used to set the environment up and load the application itself
- The `mount` function is used to mount and render the component (for example, mounting starts when the checker indicates that the previously inactive application is now active)
- The `unmount` function is used when the mounted component should be removed (for example, the unmounting is triggered when the checker indicates that the application is not active anymore)
- The `unload` function is optionally available to stop any framework-specific parts – this can only be triggered explicitly

All of these lifecycle functions should return a `Promise` to signal when the lifecycle part is completed.

In the case of popular frameworks such as React, helper libraries exist to do all that for us. For instance, using `single-spa-react` we could expose a micro frontend's component like this:

```
import React from 'react';
import ReactDOM from 'react-dom';
import rootComponent from './MyComponent.jsx';
import singleSpaReact from 'single-spa-react';

const lifecycles = singleSpaReact({
  React,
  ReactDOM,
  rootComponent,
});
```

```
export const bootstrap = lifecycles.bootstrap;
export const mount = lifecycles.mount;
export const unmount = lifecycles.unmount;
```

This enables us to focus on the actual component instead of taking time to do the technical integration.

As we can see, it's quite straightforward to declare the lifecycle using existing libraries. But even without a library, it may be rather easy. Let's consider a framework without an official library: **Solid**.

The first step is to get familiar with the framework. How does it work? Is it similar to some other framework where a lifecycle helper exists? What dependencies are needed? Can it dynamically be started and stopped? How does resource management work?

Luckily, these days most frameworks work quite similarly – at least from the outside. Furthermore, most frameworks are also happy to co-exist with other frameworks in a single DOM.

In the case of Solid, we need to use a combination of its `render` and its `createComponent` functions for mounting, while using `render` for unmounting, too. While the `bootstrap` function could lazy load the `solid-js` dependency we can also return a resolved `Promise`:

```
import { createComponent, render } from 'solid-js/dom';
import rootComponent from './my-component.js';

export const bootstrap = () => Promise.resolve();
export const mount = (props) => Promise.resolve(render(
  () =>  createComponent(rootComponent, props),
  props.domElement
));
export const unmount = (props) => Promise.resolve(render(
  () => undefined,
  props.domElement
));
```

This implements the minimum requirements for a lifecycle. Note that to make this simple wrapper a library, we should be fully asynchronous – not just provide synchronous code with a `Promise` wrapper.

Equipped with such a lifecycle, we can expose components without requiring framework compatibility. There is just one mechanism missing to use cross-framework components everywhere. Let's dive into that.

Using cross-framework components

It seems reasonable that top-level micro frontends can all be written using different frameworks, but this is not the most interesting use case. A much more interesting use case is when we use nested components, such that a page is exported from one micro frontend but this page uses components from other micro frontends – components that may have been written with different frameworks.

Let's consider the following piece of React code:

```
export const ProductPage = ({ sku }) => {
  return (
    <>
      <h1>Model Store</h1>
      <BasketInfo sku={sku} />
      <ProductInfo sku={sku} />
      <BuyButton sku={sku} />
      <Recommendations sku={sku} />
    </>
  );
};
```

So far, so good. But what if some of these components, such as `BasketInfo`, should come from other micro frontends?

One possible strategy is to get HTML placeholder elements and render directly into them. So, we change the code to have the following:

```
export const ProductPage = ({ sku }) => {
  const basketInfoRef = React.useRef(null);

  React.useEffect(() => {
    // Render the BasketInfo component when possible
    // i.e., when the container is available via the ref
    if (basketInfoRef.current) {
      return renderBasket(basketInfoRef.current, { sku });
    }
  }, []);

  return (
    <>
      <h1>Model Store</h1>
      <slot ref={basketInfoRef} />
      {/* ... */}
    </>
  );
};
```

The `renderBasket` function now would need to know which component from where to render. In single-spa, the `mountParcel` function could be used for this purpose when in an application context. Outside this context, we can always use the `mountRootParcel` function, which is directly exported from `single-spa`:

```
import { mountRootParcel } from 'single-spa';

const BasketInfo = () => import('mf-blue/basket-info');

function renderBasket(domElement, props) {
  mountRootParcel(BasketInfo, { ...props, domElement });
}
```

In this example, the `BasketInfo` value refers to the lifecycle of the component. Note that in the special case of React, the `single-spa-react` utility library already brings a convenience wrapper called *Parcel*. This way, we don't have to work with `useRef` and the DOM directly.

Generally, it makes sense to introduce a framework-agnostic broker such as `mountParcel` when dealing with cross-framework components. The great advantage is that this broker defines a single way of converting to – and from – these arbitrary components. That way, instead of having to deal with N^2 converters for N frameworks, we'd only need 2N converters – N frameworks to the broker and N frameworks from the broker.

Once established, cross-framework components are as natural as others. They are only touched from the outside, where another kind of optional lifecycle called `update` may be triggered. Most changes, however, will not happen directly via its external API, but rather from another point in the application via some communication pattern.

In the next section, we'll explore the available communication patterns between components from different micro frontends.

Exploring communication patterns

Running multiple SPA frameworks independently is a good start, but it's not fully what we'll need in real-world applications. In many applications, we will need to have some form of communication between these – at least on some occasions.

There are three quite important aspects that need to be covered by these communication requests:

- Sharing some messages, for example, to inform other micro frontends about a specific event
- Sharing data, for example, to spread knowledge efficiently about shared domain objects
- Providing components, for example, to bring `get` functionality from a different domain to a larger UI building block

All of these three aspects can be implemented following the architecture patterns discussed in the next subsections. We'll start with the exchange of events.

Exchanging events

By far the most crucial communication pattern for micro frontends is the usage of events. There is a clear reason for this: events are by default loosely coupled and quite reliable. They will not break if a handler is not registered or if the usual emitter is not there. They also properly communicate their intention of being not strictly timed.

In a nutshell, there are many ways to set up events for micro frontends. The most common, potentially the most reliable, and definitely the best-performing way is to leverage the DOM API. The DOM API offers so-called custom events, which give us the ability to use the standard event listener system with our own names and payloads.

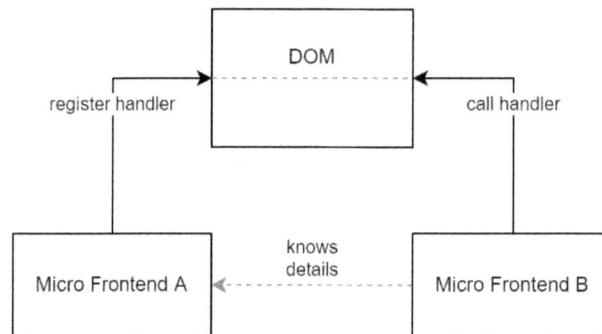

Figure 10.4 – The standard DOM event interface can be used to exchange events

In the preceding figure, we see the distilled architecture. There is some implied knowledge between the micro frontends emitting and reacting to a particular event. After all, there will always be something known – even in loosely coupled strategies.

Using the custom event API is quite straightforward. For the event emitter, this looks like this:

```
window.dispatchEvent(new CustomEvent('my-event', {
  detail: {
    message: 'Hello!',
  },
}));
```

Listening for this custom event requires only a bit of code:

```
window.addEventListener('my-event', e => {
    // use e.detail.message
});
```

The `detail` property is where custom data is attached. Sometimes, it could be useful to place a custom API on top of custom events. In practice, this could be played such that the DOM API is only used as transport mechanism, with a custom API being used to add or remove listeners and emit events.

One of the things that events cannot cover well is guaranteed access to shared data.

Sharing data

It may not be as important as events, but in many scenarios, the right call is to define how the same set of data should be read and written.

Like before, we need to provide a central location for accessing the shared information. This could be the `window` object, a shared dependency, or a handle provided by some other reachable object.

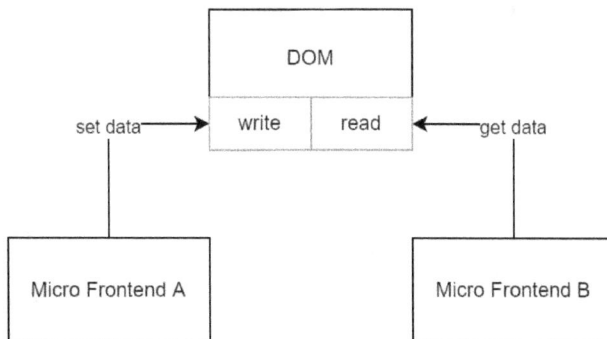

Figure 10.5 – A central API provides read and write access to shared data

The architecture of this communication pattern is illustrated in *Figure 10.5*. Importantly, we require two separate APIs – one to write new data and one to read existing data. This API could define a concept of ownership, which could be used with different kinds of access privileges. For instance, only the owner of some data could be allowed to write it. In this setup, reading could be done by any micro frontend.

In most cases, a shared data store is most likely an indicator of an underlying issue. Quite often, the domain is not properly split. It is highly probable that this is due to the lack of technical ability to extend components with UI fragments from other micro frontends – in a loosely coupled form. Let's see how this could be brought in, though.

Extending components

Almost the most important communication pattern is to define a way for existing components to be extended with more functionality coming from other micro frontends. In general, this is a big problem: Consider a table with data coming from one micro frontend – such as the transaction table we've seen in the example. How can one micro frontend add additional value to this table?

As an example, let's say that a micro frontend could infer tax-deductible transactions. Wouldn't that be nice information that should be part of that table? Clearly, having this information (optionally) within the existing table would be great. The one thing that's missing is how to do it. The obvious way would be to either integrate that into the transaction table directly or import a component from another micro frontend. Neither option is really ideal.

In the first option, we violate the domain decomposition. While technically fine, this will present a challenge in terms of team knowledge and maintenance. However, the other option is even worse. Here, we do not only violate the domain decomposition, we also introduce strong coupling to another micro frontend.

To solve this issue, we can create a so-called **extension slot**. This allows additional components from any micro frontend to enter an existing component. In the following figure, the general flow is outlined.

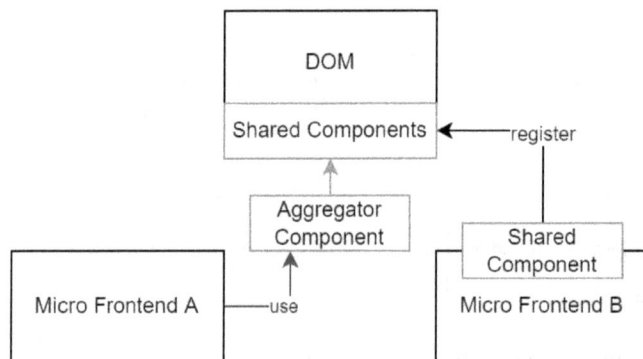

Figure 10.6 – Using a neutral aggregator component to render all components for an extension slot

The extension slot is represented by a special aggregator component. The responsibility of this component is to display the components registered for the slot. This way, the whole pattern is a bit like events – just for UI components.

While events work with an emitter and multiple handlers to listen for emitted events, extension slots work with a renderer (extension slot) and multiple components to show up when being rendered. Like with events, an object with properties describing the current rendering is passed to the components.

While loose coupling and component scalability are important for micro frontend solutions, one aspect that should not be skipped is to care about an application's performance. Using micro frontends can lead to sub-optimal performance, however, due to the increased modularity the performance might actually even be better than in monolithic applications.

One area where performance can really be optimized is by applying modern hydration and progressive rendering techniques.

Optimizing hydration and progressive rendering

One thing that is elementary in SPAs, but also one of the problems of using SPAs, is that they are very heavy on JavaScript – they are essentially just scripted applications. This is especially a problem for e-commerce websites and content-heavy pages. Here, the resulting bundle size and the application load time might be just not good enough to justify writing a SPA in the first place.

The problem with load times does not get better when we introduce micro frontends using SPA composition. With this addition, the performance might actually be worse. However, looking at the techniques to improve SPA performance also yields the key to improve the performance of micro frontends using the SPA composition.

First, let's see what we can do to improve performance initially. We can reduce JavaScript. That works up to a certain point. Instead of loading all the JavaScript of the application immediately, we only load the parts required to display the application's start screen. Everything else will be lazily loaded, i.e., on-demand only when we need it.

Splitting the application up in multiple bundles is key to reduce the JavaScript payload for a SPA. The same concept also works for micro frontends. If no components from a micro frontend are needed then the respective micro frontend should not be loaded. Likewise, only the components that are currently needed should be loaded.

While the bundle splitting above has some advantages and is generally a good approach, it does not yield the best possible performance. After all, we still need to wait until the core JavaScript chunks are loaded and rendered before the application is usable. Instead, we want to display something right away. This is where techniques such as hydration and progressive rendering come in. Both rely on server-side rendering to optimize the initial loading time.

Hydration is the name for making the rendered DOM interactive. It works by first rendering the website completely on the backend, i.e., spitting out HTML that is then delivered to the client. The transmitted HTML would be static – there are no event handlers or any JavaScript logic connected to it yet. To make this static HTML interactive a special kind of rendering is performed on the client – the so-called hydration.

By using hydration instead of a normal rendering, the information from the backend is used to continue where the server left off. This way the performance is supposed to be better and the load time is faster. The concept is summarized in *Figure 10.7*.

Rendered HTML   Load JavaScript   Hydrate

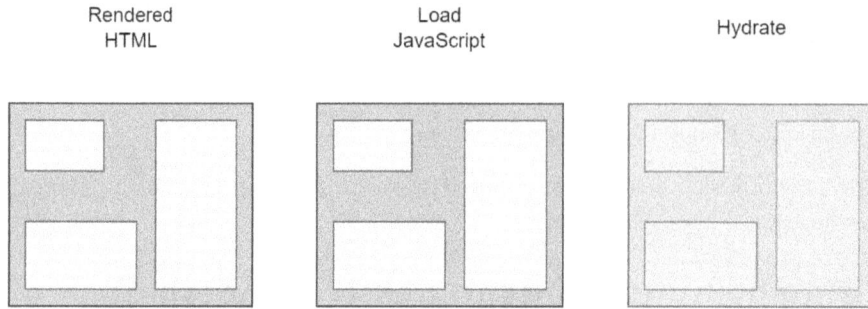

Figure 10.7 – The concept of hydration for optimizing SPA performance

The backend performs an initial server-side rendering resulting in HTML that is transported to the client. The HTML contains a reference to the script representing the SPA – just as in case of a full SPA without server-side rendering. Finally, once the JavaScript is loaded, the hydration can kick in and make the whole page interactive.

While improved performance is certainly true in most cases it is certainly not a silver bullet. And even if better performance is achieved, the main problem is still that interactivity requires loading the whole JavaScript – even though hydration should still be using bundle splitting and thus contain reduced script sizes. In any case, making a "full hydration" is wasteful. A much better alternative is progressive rendering.

Progressive rendering does not make a hydration on the full tree, but only of the parts that require interactivity. This is usually one part of the islands architecture implementation as introduced in *Chapter 7*. For micro frontends, however, we get this essentially for free on a component level.

To illustrate the difference to plain hydration we can have a look at the general principle outlined in *Figure 10.8*.

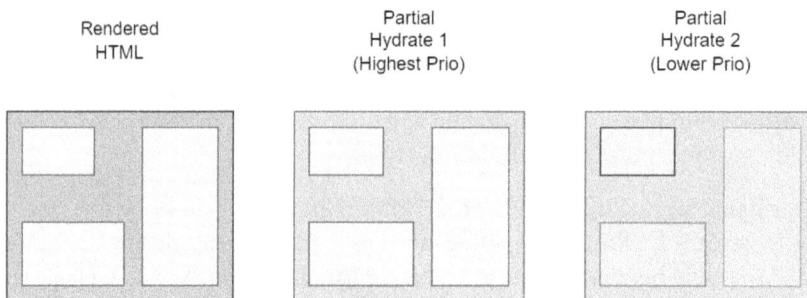

Rendered HTML   Partial Hydrate 1 (Highest Prio)   Partial Hydrate 2 (Lower Prio)

Figure 10.8 – The concept of using progressive rendering for performance optimization

There are two crucial properties here; the first is that partial hydration can kick in much faster than pure hydration – mostly because the highest priority scripts should be inlined and transmitted directly with the rendered HTML. The second is that components are hydrated independently from each other. This results in a state where some components might be fully hydrated, while others are still waiting to be hydrated – or skip hydration altogether.

The implementation of progressive hydration usually favors JavaScript or even Node.js-based frameworks. This way, the standard developer experience of a SPA is preserved, while the framework decides which parts of the application are rendered on the server, on the client, or on both. Most frameworks will have a default way – usually rendering on the server – that can then be changed by the developer.

For micro frontends to fully leverage progressive hydration, we basically require a mix consisting of a server-side composition with a SPA composition. We'll look at an example implementation using the Qwik framework at the end of *Chapter 11*.

Summary

In this chapter, you learned how to compose multiple single-page applications into one coherent web application. You saw that user interaction is meant to be quite high, however, to achieve a great user experience, a lot of technical challenges need to be tackled.

By applying the communication patterns covered in this chapter, you can ensure loose coupling. This helps your application to scale as you go. In the end, the goal is always to exchange only the minimum amount necessary for integration.

In the next chapter, we take this pattern a bit further and create a system that mirrors serverless functions just for composing the UI. This should give us a system that is both very flexible and simple to develop while maintaining a great, coherent user experience.

11
Siteless UIs

In the previous chapter, we saw how the SPA composition pattern can be used to combine multiple SPAs into one unified application. While this pattern is attractive due to a variety of business and technical factors, we've also drilled deep enough to observe its challenges.

A pattern that directly builds on top of the SPA composition is the creation of siteless UIs. In siteless UIs, serverless concepts such as modularity, dynamic loading, and runtime-driven execution are applied to frontends. Micro frontends can be dynamically discovered and integrated at runtime without the app shell needing to know their specifics in advance, enhancing flexibility and reducing coupling.

In this chapter, we'll close our tour of the different micro frontend patterns by taking a deep dive into next-generation micro frontends with the siteless UI pattern.

We'll cover the following topics in this chapter:

- Basics of siteless UIs
- Advantages and disadvantages of siteless UIs
- Comparing siteless UIs and serverless
- Creating a siteless UI runtime
- Writing siteless UI modules
- Implementing islands composition with Qwik

This should help us understand when the siteless UI pattern makes sense and how we can implement it practically.

Technical requirements

To follow the code samples in this book, you need knowledge of JavaScript. You also need to know how to use the command line. You should have Node.js (version 20 or higher) installed using the instructions at `https://nodejs.org`. For this chapter, additional knowledge of using the webpack bundler can be helpful.

The code used in this chapter can be found in the following repo:

`https://github.com/PacktPublishing/The-Art-of-Micro-Frontends-Second-Edition/tree/main/Chapter11`

The CiA video for this chapter can be found at `https://packt.link/GZshi`

The basics of siteless UIs

The siteless UI pattern builds directly on top of the SPA composition. It leverages a plugin architecture, which has been known for its flexibility without the loss of consistency. The plugin architecture in siteless UIs allows micro frontends to be developed and integrated as independent modules. Each module can be loaded dynamically and executed within a shared runtime environment, ensuring consistent behavior across the application while allowing individual teams to work independently on different parts of the frontend.

Like in the previous chapters, we will also introduce this pattern using our standard structure. We will start by taking a closer look at its architecture before we try to come up with a simple sample implementation. Finally, we will discuss potential enhancements of the sample.

Let's start with a closer look at its architecture.

The architecture

One of the issues with the SPA composition pattern was that the app shell was a bit too constrained, in the sense that the app shell most likely knows where all micro frontends reside and how to integrate them. This kind of tight coupling will become problematic quite fast as changes to the micro frontend solution's infrastructure, such as the addition of a new micro frontend, usually require a change in the app shell as well. Optimistically, we'd like to avoid this kind of tight coupling to truly empower teams and make the whole application easier to maintain.

In contrast, in an evolved version, the app shell itself would not know about the micro frontends. Instead, it would rely on some other part to identify which micro frontends to resolve and integrate. The integration itself would be left to the micro frontends.

The following diagram illustrates the core architecture principles of the siteless UI pattern:

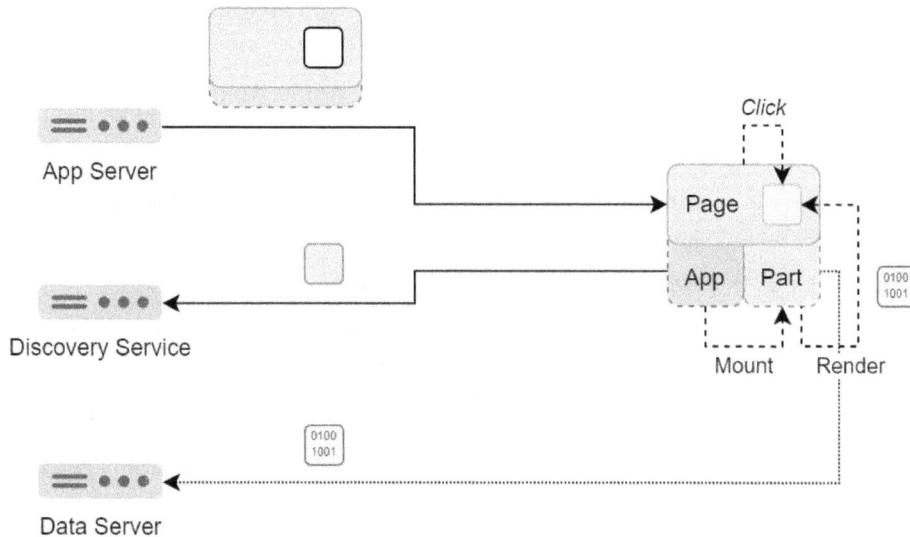

Figure 11.1 – The siteless UI pattern

In *Figure 11.1*, we can also see the similarities to the SPA composition. What makes the siteless UI pattern unique is that it naturally decouples the micro frontends. It does that by using a central source where the micro frontends are published and stored. The information regarding which micro frontends are relevant for the current application is called a **feed of micro frontends**. The feed is retrieved from a dedicated service known as a micro frontend discovery service.

Furthermore, instead of being fully autonomous, micro frontends are modules that are still run within the execution context of the app shell. To provide these kinds of modules, we can use a plugin architecture. One way of introducing a plugin architecture is to define two things:

- A general life cycle of the modules exposed via functions
- An API to be handed over to the plugins that is accessible during their life cycle

In the simplest approach, we reduce the life cycle to a single step: the setup of a plugin. In this case, each module exposes a single function, for example, called `setup`. The app shell can now call these setup functions, providing an API object as an argument to the function. The content of the object is left to the app shell owner. It makes sense to provide dedicated methods, which allow the registration of components for specific parts of the UI such as a navigation or content area, here.

Using the exports of the scripts is a clear change in direction when compared to the SPA composition beforehand. Instead of using global variables, we now work exclusively in modules. While global variables may still be used somewhere under the hood, such as to get some singleton behavior or share some dependencies, we should never see or interact with these global variables directly.

Let's try to experience the siteless UI pattern by modifying the sample from the previous chapter.

Sample implementation

To illustrate the workings of this pattern in the most extreme setup, we'll distribute all parts of the solution. This means we'll end up with five repositories:

- One repository for the micro frontend discovery service providing the feed of micro frontends – we'll just refer to this as the **feed server**

- One repository for the app shell orchestrating the micro frontends

- One repository for the tax micro frontend providing a tax-deductible information component for the balance table (or other interested micro frontends)

- One repository for the settings micro frontend providing the user settings page

- One repository for the balance micro frontend providing the dashboard page with the balance sheet

In the following sections, we'll cover each repository. We will start with the feed server.

Feed server

The feed server can be implemented as a simple Node.js application using Express. The core part of the server consists of three different route handlers, which can be defined like this:

```
app.get('/modules', getLatestModules());
app.post('/modules', publishModule(host));
app.get('/files(/@:org)?/:name/:version/:file?',
  getFiles());
```

The first route handles all the requests for retrieving the available modules. It will respond with the latest modules.

Even though our feed server may store multiple versions for each micro frontend, we'll only serve the latest one in the sample. With further APIs or efforts, this enables quick rollback or A/B testing scenarios, as follows:

```
exports.getLatestModules = () => async (_, res) => {
  const modules = await getAllModulesFromDatabase();
  const unique = modules.reduce((prev, curr) => {
    prev[curr.meta.name] = curr.meta;
    return prev;
  }, {});
  const items = Object.keys(unique).map((name) =>
```

```
    unique[name]);
  return res.json(items);
};
```

There are multiple ways of publishing modules. One way could be to couple to some other service – for example, providing a private npm registry.

A more integrated way would be to allow the upload using a form with a file entry. An implementation of this approach is as follows:

```
exports.publishModule = (rootUrl) => (req, res) => {
  const bb = req.busboy;
  req.pipe(bb);

  bb.on('file', async (_, file) => {
    try {
      const content = await getModuleContent(file,
        rootUrl);
      await setModuleData(content);
      res.sendStatus(200);
    } catch (err) {
      res.sendStatus(400);
    }
  });
};
```

The getModuleContent function is used to extract the content from the provided file. In our example, we'll use an npm package, which is just a tgz file. We'll discuss in the *Publishing modules* section why this might be a good choice.

Finally, we need a way to retrieve the files. Ideally, this will not be part of the micro frontend discovery service. However, in this ideal scenario, we would need another service – most likely a kind of **CDN** with access to really fast storage. Therefore, we'll just provide an endpoint that is capable of serving any kind of demanded file – if it was part of a previously uploaded micro frontend.

Now, we will discuss the function to create the request handler for these files. The thing we need to be careful about is the potential naming of the packages. Since npm packages can be scoped, a single path separator may appear in the URL part, representing the package name. We need to apply some logic to correctly identify this part, which is shown in the following code:

```
exports.getFiles = () => async (req, res) => {
  const { name, version, org, file } = req.params;
  const id = org ? `@${org}/${name}` : name;
  const moduleData = await getModuleData(id, version);
```

```
  if (moduleData && file) {
    const path = path.join(moduleData.root, file);
    const content = Buffer.from(moduleData.files[path]);

    if (content) {
      const tenYears = 24 * 60 * 60 * 365 * 10;
      return res
        .header('Cache-Control', `public, max-
          age=${tenYears}`)
        .contentType(mime.lookup(file) ||
          'application/octet-stream')
        .status(200)
        .send(content);
    }
  }
  res.status(404).send('File not found!');
};
```

With these handlers implemented, we can start the server and see whether everything works fine. We'll keep it running for now to test the integration with the other pieces. Let's implement the app shell before we continue with the individual micro frontends.

The code for this section can be found at: `https://github.com/PacktPublishing/The-Art-of-Micro-Frontends-Second-Edition/tree/main/Chapter11/11-siteless-uis/service-feed`

App shell

The app shell in the siteless UI pattern acts as the core orchestrator that initializes the application, retrieves the list of available micro frontends from the feed server, and dynamically loads and integrates them into the application at runtime. It provides a consistent runtime environment, as well as the necessary APIs for micro frontends to interact seamlessly. This looks quite similar to the script loading from the SPA composition sample (which we saw in *Chapter 10*).

However, in contrast to the example from *Chapter 10*, we are not loading a static file. Also, we need to do a bit more work than before. Unlike the SPA composition pattern, where each micro frontend reaches out to some functionality, we need to provide a set of APIs in a well-defined interface.

In the following example, we assume that each micro frontend will be exposed as a global variable yielding its exports. In these exports, we'll look for a `setup` function. The function is finally called by the app shell providing the API object for the micro frontend to use, as follows:

```
const feedUrl = "http://localhost:9000/modules";

fetch(feedUrl).then(res => res.json()).then((modules) =>
  modules.forEach((moduleData) => {
    const script = document.createElement("script");
    script.src = moduleData.link;
    script.onload = () => {
      const nsName = moduleData.name;
      const { setup } = window[nsName] || {};

      if (typeof setup === "function") {
        const api = createApi(nsName);
        setup(api);
      }
    };
    document.body.appendChild(script);
  })
);
```

Keep in mind that the URL of the feed server should be adapted to our case. For running the sample locally, the given URL may be adequate. However, in general, we should make the public base URL configurable, allowing the use of a different URL for our production environment.

We should create an API object per micro frontend. The advantage of such an approach is to avoid potential conflicts and manipulations between the different micro frontends.

For this sample, we'll provide a simple yet effective API, pretty much providing the same functionality as before.

One advantage of creating an API object per micro frontend is that we can immediately recognize that we don't have to specify the name of the micro frontend in all the APIs. This can be captured when creating the APIs. Since we will do that per micro frontend, we can also capture micro frontend-specific information.

The general outline for the API creation can be summarized as follows:

```
function createApi(nsName) {
  return {
    registerPage(route, lifecycle) {
      // ...
    },
```

```
    registerExtension(id, lifecycle) {
      // ...
    },
    on(eventName, handler) {
      // ...
    },
    off(eventName, handler) {
      // ...
    },
    renderExtension(container, id, props) {
      // ...
    },
  };
}
```

For instance, the `registerPage` API could use a combination of the `registerComponent` and `activateOnUrlChange` functions, which we've seen in the sample for the SPA composition.

When applied with the captured information, the implementation may be as simple as this:

```
registerComponent(nsName, route, lifecycle);
activateOnUrlChange(
  nsName,
  route,
  (location) => location.pathname === route
);
```

We don't need to change the build process. Like before, we are free to choose whatever tooling we want. In this case, we will keep the Parcel bundler for building and development.

The code for this section can be found at: `https://github.com/PacktPublishing/The-Art-of-Micro-Frontends-Second-Edition/tree/main/Chapter11/11-siteless-uis/app-shell`

With the app shell up and running, we can focus on the different micro frontends. We'll start with the micro frontend concerned with tax information.

Micro frontend – tax

The `tax` micro frontend does not require many changes. The `Info` component itself can still be provided as a Svelte component. There are no changes needed to this component.

The minimal changes required for micro frontends when transitioning to the siteless UI pattern signify the pattern's flexibility and ease of adoption. It demonstrates that existing micro frontends can be integrated into the new architecture with little to no modifications, ensuring a smooth migration and preserving previous development efforts.

The root module of the micro frontend requires some changes. It now reads as follows:

```
let Info = undefined;

export function setup(api) {
  api.registerExtension('balance-info', {
    bootstrap: () =>
      import('./Info.svelte').then((content) => {
        Info = content.default;
      }),
    mount: (target, props) =>
      new Info({
        target,
        props,
      }),
    unmount: (_, info) => info.$destroy(),
  });
}
```

We've transported the life cycle defined earlier in the app shell into the new way. This way leverages an exported setup function that receives an object containing the app shell's API.

What we see is that the micro frontend's name does not need to be mentioned when we call the registerExtension function. What we changed here is to use a more appropriate name. We'll see that loose coupling demands giving careful names to the extension slots instead of directly obtaining components via the names of their micro frontends.

The code for this section can be found at: https://github.com/PacktPublishing/The-Art-of-Micro-Frontends-Second-Edition/tree/main/Chapter11/11-siteless-uis/frontend-tax

Let's explore the settings micro frontend to review more of the changes that need to be applied.

Micro frontend – settings

Again, the major changes are within the root module. However, we need to be careful with the package.json file too. Since we will use the information stored in this file on our feed server, we need to have accurate information in there.

Previously, we did not keep the meta information regarding the `main` field accurate. However, this should reflect where the `build` artifact is placed. In our case, we will choose `dist/index.js`. To fully accommodate our choice of the `main` field, we need to also modify the configuration for webpack.

In the case of the `settings` micro frontend, the webpack configuration reads as follows:

```
const { name } = require('./package.json');
module.exports = {
  entry: {
    [name]: './src/index.js',
  },
  output: {
    filename: 'index.js',
    library: name,
    path: __dirname + '/dist',
  },
  // ... as beforehand
};
```

We have reused the information from `package.json` to create a consistent and reusable configuration file. The output filename is now fixed to `index.js`. Furthermore, we have used the library target with the same name. This follows the convention established earlier in the app shell.

Consequently, the build process has remained very similar, but it now handles the exports via a global variable. The necessary wrapper is introduced by webpack.

The code for this section can be found at: `https://github.com/PacktPublishing/The-Art-of-Micro-Frontends-Second-Edition/tree/main/Chapter11/11-siteless-uis/frontend-settings`

With these changes in mind, let's look at the most important micro frontend: the one hosting the balance sheet.

Micro frontend – balance

The `balance` micro frontend also needs to have its metadata, webpack configuration, and root module modified. Additionally, we also need to apply the new way of rendering the extension components registered in the app shell.

What we'll do is provide the `renderExtension` API to the `BalanceSheet` component. In the root module, this looks as follows:

```
export function setup(api) {
  api.registerPage('/', {
    mount: (target) => render(
      <BalanceSheet onRender={api.renderExtension} />,
```

```
        target
    ), // ... remaining lifecycle as beforehand
  });
}
```

This way, the component can also use the `renderExtension` API. In our case, we could just provide the `onRender` prop to the interested children, who may provide that prop to their children again.

> **Important note**
>
> In React, there are two preferred ways of passing in values. One is via props, while the other is via context. Usually, we would prefer props. However, under certain circumstances, context may be even better. One situation where contexts shine is when we otherwise would see a phenomenon known as **prop drilling**. This appears when we need to pass on the same prop multiple levels down just to give a fairly distant child element a prop.

Now that we have transformed all of these micro frontends, it's time to publish them on the feed server. This means that first, we need to build and pack them as follows:

```
npm run bundle
npm pack
```

For the actual upload, we can use a graphical tool such as Postman or a command-line utility such as **cURL**.

As an example, the balance micro frontend in version 1.0.0 is published with the following command:

```
curl -F 'file=@./balance-1.0.0.tgz' http://localhost:9000/modules
```

After all the micro frontends have been published, the feed server's response is no longer empty. We should see a response similar to the following snippet:

```
[
  {
    "name": "balance",
    "version": "1.0.0",
    "link": "http://localhost:9000/files/balance/
      1.0.0/index.js"
  },
  {
    "name": "tax",
    "version": "1.0.0",
    "link": "http://localhost:9000/files/tax/
      1.0.0/index.js"
  },
```

```
{
    "name": "settings",
    "version": "1.0.0",
    "link": "http://localhost:9000/files/settings/
        1.0.0/index.js"
  }
]
```

So far, so good.

The code for this section can be found at: `https://github.com/PacktPublishing/The-Art-of-Micro-Frontends-Second-Edition/tree/main/Chapter11/11-siteless-uis/frontend-balance`

As usual, the question now is: what could we have done better?

Potential enhancements

Even though the solution already works quite well, it is only usable within the boundaries given by this particular sample. One reason for this is that there is almost no error handling. One thing that will happen when such an application is released to a larger audience is that problems will occur – problems we've never thought of. HTTP responses will be cut, requests will fail, and browsers that are either outdated or have been modified by company policies or installed extensions will interfere with the site. We need to be prepared for and handle such unexpected behavior gracefully.

Another thing that becomes tedious is the lack of dedicated tooling. While it's certainly appreciated to be able to decide what tooling to use per micro frontend, we still want to have some things figured out. One example is the upload to the feed server. Using cURL may be an appropriate solution on macOS and Linux, but it's not ideal on Windows.

Furthermore, with the current implementation, we cannot see the bugs until a micro frontend is published. It is like working in the dark. Granted, this was the situation with the SPA composition sample too, but it's definitely one of the first things that should be changed.

Another thing to consider is that the implementation of the feed server that we have seen is wide open. There is no restriction to stop uploads of files that are too large or come from an untrusted source. Also, existing uploads can just be overwritten – which is a huge problem with regards to the long caching that is desired.

Certainly, while the sample implementation leaves many things to be desired, the general direction is great. Compared to the sample implementation of the SPA composition, the loose coupling alone makes the solution much more resilient. Nevertheless, to give it the final touches, a lot more is needed. Later in this chapter, we'll introduce a framework to handle these missing pieces for us.

For now, let's look at the general advantages and disadvantages of the siteless UI pattern.

Advantages and disadvantages of siteless UIs

Like with the other patterns, there are also clear reasons to go for this style or to avoid it. Primarily, siteless UIs will be used in highly interactive SPAs. As such, many of the advantages and disadvantages of the SPA composition are shared. However, due to the focus on developer experience and loose coupling, the development of individual modules is much simpler. The drawback is that the app shell development is much more complex. This is a standard tradeoff – a bit of complexity for all teams is exchanged for a lot more complexity for a single team.

The greatest disadvantage of this pattern is the implied – and strong – dependency of the modules on the API provided by the app shell. If we just change one of the APIs, we might crash all modules using that API. Even worse, we might not even recognize this problem until we release the updated app shell. As the modules are deployed independently, the whole system only comes together at runtime. At compile time, such issues are not easy to spot and will usually be missed.

You may argue that most drawbacks are less related to the siteless UI pattern and more of an issue with loose coupling in general. While this is true, since the pattern embraces loose coupling, all of these challenges can now be found in siteless UI implementations too. On the other hand, we'll also find all the advantages that come with loose coupling.

Potentially the greatest advantage of the siteless UI pattern is that it pretty much guarantees a consistent application out of the box. By providing an API to access essential app functionality, all modules need to go through the same interface. Naturally, the same patterns and practices are already fully determined at this boundary. The UX is formed around the layouts, which are then communicated properly via the API. Here, this architectural style really ties all micro frontends together without crippling their freedom too much.

While big tech companies such as Microsoft use siteless UIs in some form or another already, the pattern also appeals to more classical companies with a drive for digital transformation. One example of this is ZEISS, which used the pattern to establish its new customer-facing portal. We'll also discuss real-world siteless UI implementations in greater detail later (see *Chapter 20*).

The approach that is introduced by the siteless UI pattern on the frontend is quite similar to the serverless function architecture in the backend. To fully understand why these two approaches are similar and what advantages and disadvantages they may share, it makes sense to compare them in detail.

Comparing siteless UIs and serverless

After microservices had been established on the backend, a new kind of architecture was introduced that became popular: serverless or **FaaS** architecture. In a nutshell, this architecture reduced backend services to single functions where all the essentials had been handled by an integrated runtime.

In the beginning, serverless was mainly a selling point for cloud providers. They advocated the new pattern with fewer dedicated costs. After all, since these functions use a shared and provided runtime, they are not required to run in custom containers that need dedicated resources. Instead, they can just sit idle and wait for a request, being invoked only when required. The runtime would be able to serve many different functions from many different tenants.

In the following sections, we'll focus our comparison on two major aspects: how local development works and how modules are published. We'll start with the setup for local development.

Developing locally

Almost all FaaS solutions can be developed using a standard IDE, even though many offer dedicated online IDEs. Sometimes, cloud providers offer a direct way of manipulating the code for a serverless function. For instance, Microsoft Azure was one of the first users to directly integrate the Monaco editor, which forms the basis of the popular Visual Studio Code text editor.

One of the major principles of a serverless project is that local development is not only possible but is as simple as development for a standard project. Thus, developers should be able to clone FaaS projects to build, debug, and publish them from their local machines.

With siteless UIs, we want to be able to provide the same experience. A new developer has to be able to clone the code of a single micro frontend and start a local debugging session. It should not be necessary to know any URLs or special tricks to inject the module into an instance running in a special development environment. Instead, the whole experience has to be as local as possible.

By using an emulator package, this is possible. The emulator is usually just a special development build of the runtime provided within the app shell. Since it runs the same code, it will work like the original. The downside of this approach is that the emulator will require some installation. While the npm infrastructure helps us here, an installation also assumes that a specific version is picked. Consequently, the likelihood of dealing with an outdated version of the emulator is quite high.

Using outdated versions of the emulator does not sound like a problem at first. There are, however, some immediate issues spawning from an outdated installation. First, the functionality may have changed, thus giving us wrong typing information and leading to runtime errors later. Second, the design that we see on the screen may have changed too, which may lead to design inconsistencies when released.

A fix for most problems is to ensure that proper end-to-end testing is done after deployment. Also, insisting on doing a more extensive integration check is part of the solution. In general, none of these should be new things. Thoughtful quality assurance and detailed validation of acceptance criteria are integral parts of the development process in many projects.

In the end, most checks can be done when actually publishing the modules anyway.

Publishing modules

Every FaaS platform requires a way to quickly have existing functions updated or new functions added. While microservices leverage container formats such as Docker, FaaS usually leverages simple archive formats such as `tar` files.

For micro frontends, the situation is similar. Backend-driven micro frontends are usually delivered in container formats – but even client-side composed variants may be released that way. For siteless UIs, a simple archive format is sufficient. A popular choice may be a zipped tarball using the `tgz` extension.

There are multiple advantages to using `tgz` files as the package format. The most important point is that the format is already produced and used by the `npm` tool itself. Being compliant with the most important development tooling is huge with respect to developer productivity. Instead of relying on custom tooling, standard applications can be reused.

The npm package format also gives us a predefined location for metadata: the `package.json` file. This file also helps us get some information about the files contained in the package. For instance, the value of the `main` field tells us the location of the entry module representing the package.

While the FaaS offering of a cloud provider comes with an integrated way of accepting new functions, in a custom siteless UI solution, we need to provide a service to handle that. Usually, this responsibility falls to the feed server.

> **Important note**
>
> An open source implementation of a feed server is the Piral Feed Service. The available source code allows publishing and provisioning modules provided in an npm package. A cloud version of the service exists as a free community edition, as well as a commercial enterprise offering packaged as a Docker container. These versions have more features such as dynamic module provisioning. More information can be found at `https://www.piral.cloud/`.

The feed server is therefore a core infrastructure piece that provides a way of writing micro frontends without having to take care of the core infrastructure such as a deployment target. If it is taken from an existing SaaS offering, we can build a frontend solution that does not require any infrastructure at all, as everything can be either hosted on some static storage or provided via a third-party cloud service.

All this makes siteless UIs quite similar to serverless functions – just working at the frontend level instead of the backend. As we saw, the most important piece in this construct is a runtime. Let's explore in a bit more depth what needs to be considered when implementing the siteless UI pattern.

Creating a siteless UI runtime

We've already seen that a runtime forms the centerpiece of a siteless UI implementation. While the runtime forms as an orchestrator when released and running in production, it also provides an emulator to allow for local development of new modules.

This way, things such as automatic provisioning, caching rules, and runtime optimizations are all available in production – but they don't reduce developer efficiency during development.

In order to focus on the decisions that matter for our runtime, we should pick an established framework to provide the technical basis. One possible choice is Piral.

> **Important note**
>
> Coming up with a full siteless UI implementation is difficult. It requires not only a model for the app shell that emphasizes local development but also that the feed server and micro frontend packaging are defined and implemented while adding a custom API with reliability for micro frontends. A quick way around this is to use the Piral framework, which is essentially a tool for creating plugin-based UIs, thus providing everything needed for a siteless UI implementation.

In the following sections, we will look into the creation and deployment of an app shell with Piral. We will start by building a runtime before actually deploying it.

Building a runtime with Piral

Like many other frameworks, Piral comes with many templates, boilerplate generators, and tutorials to get started quickly. We will not use the standard template and will instead start by creating a new project from scratch.

The first command is to create a new `package.json` file using the npm command-line tool:

```
npm init -y
```

Then we can install the dependencies. In this case, we'll use the `piral` package and the `piral-lazy` plugin to represent the runtime, while `piral-cli` and the `piral-cli-webpack` package are used to actually build it.

The commands to install these dependencies are as follows:

```
npm i piral piral-lazy --save
npm i piral-cli piral-cli-webpack5 --save-dev
```

The app shell is transported to the client in the form of an HTML file. To tell the Piral CLI where the HTML file sits, we need to change `package.json`. We'll add this:

```
{
    ... // previous properties
    "app": "src/index.html"
}
```

The HTML file may look as simple and elementary as this:

```
<!DOCTYPE html>
<html lang="en">
<head>
<meta charset="UTF-8">
<title>My Siteless UI App</title>
</head>
<body>
<div id="app"></div>
<script src="./index.jsx"></script>
</body>
</html>
```

The reference to the loader script is placed beside standard HTML templating. This is used by the tooling to define the entry point of the bundler. We may also directly reference TypeScript files here, as the HTML is only used to identify the sources. The sources will be replaced with their generated filenames in the build artifacts.

The loader script can then be written with a simple structure in mind, as follows:

```
import * as React from 'react';
import { createRoot } from 'react-dom/client';
import { createInstance, Piral } from 'piral';
import { createLazyApi } from 'piral-lazy';
import { createCustomApi } from './api';
import { errorComponents, components } from './components';

const feedUrl = 'https://feed.piral.cloud/api/v1/
  pilet/<feed-name>'; // replace with URL to your feed

function requestPilets() {
  return fetch(feedUrl)
    .then(res => res.json())
    .then(res => res.items);
}

const instance = createInstance({
  requestPilets,
  state: { errorComponents, components },
  plugins: [createCustomApi(), createLazyApi()],
});

const root = createRoot(document.querySelector('#app'));
root.render(<Piral instance={instance} />);
```

In Piral's terminology, the modules are called **pilets**. In the loader, we need to give Piral a function that is able to connect to a feed server responding with a list of pilets. The function for specifying this is called requestPilets. In this example, we will use the free community version of the Piral Feed Service to host the pilets. Make sure to replace the feed-name placeholder with the name of your feed.

The integration of the piral-lazy plugin package was done by adding the result of calling the createLazyApi setup function to the provided plugins. The components and errorComponents objects are filled with components to define the display elements for common layout and error parts of the UI.

The api module could be defined like this:

```
export function createCustomApi() {
  return context => ({
    myApiFn() {
      console.log('Hello from the API!');
    },
  });
}
```

This would add a myApiFn function to the API available in the modules. At this time, we can use the Piral CLI to bundle and deploy our runtime.

Deploying a runtime with Piral

Building an app shell is already integrated in the Piral CLI. The build command automatically processes the package.json file of the current folder and processes the HTML file there.

All we need to do is run the command (for example, via the npx task runner), as follows:

```
npx piral build
```

This will result in two artifacts. One is a dist/release folder that contains files, which can be uploaded to some static storage. This is the production build of our app shell. The dist/emulator folder only contains a tgz file. This is the emulator bundled together in an npm package, which could now be published on the official registry or some private npm registry.

Using GitHub as a basis for our runtime, we can also leverage GitHub pages for the simplified hosting of static websites. The simplest way to follow the right steps is to install the gh-pages package as follows:

```
npx gh-pages -d dist/release
```

This will take all the files from the dist/release folder and put them in the root directory of the gh-pages branch. Usually, this is an orphan branch, which means a branch that is not based on some other branch and contains its own history of changes.

Pushing this branch leads to a new site, available at `https://<username>.github.io/<project>`, where `username` represents your GitHub username and `project` represents the name of the repository. The website should be live within seconds.

Afterward, we can publish the npm package representing the emulator. The easiest way is to use the npm command-line utility as follows:

```
npm publish dist/emulator/<package-name>-<package-version>.tgz
```

Importantly, for `npm publish` to work, we need to be authenticated using our `npmjs.com` credentials. Otherwise, an error (`ENEEDAUTH`) is displayed. To authenticate the npm CLI tool, the `npm adduser` command needs to be used. Additionally, the chosen package name has to be unique and not already owned by somebody else.

Now that we have a runtime that is not only live and online but also available in the form of an emulator, it's time to write some modules for it.

Writing siteless UI modules

We've already heard that a module for an app shell created with Piral is called a pilet. Pilets are developed like most frontend applications – with the small exception that pilets are not self-contained applications, but closer to independent libraries.

If we want to start a new pilet, we have two options: we can use the integrated tooling to scaffold a new project with the right setup, or we can start from scratch as we did with the app shell. Once we have published the emulator, we can scaffold a new project from the command line:

```
npm init pilet -- --source <package-name> --bundler webpack5
--defaults
```

This will initialize a new npm project with a `package.json` file representing a pilet. The pilet will be targeted at the previously published emulator package. For bundling purposes, we will use `webpack` again. If a private npm registry was used for publishing the emulator, the URL in the command needs to be adjusted too.

Once we have scaffolded a new pilet, we can start development right away. We'll continue our investigation by first going into detail on the life cycle of a pilet before we see how framework-agnostic components can be implemented in such a setup.

Looking at a pilet's life cycle

When we discuss the life cycle of a pilet, we need to consider two different kinds of life cycles:

- The software development life cycle of a micro frontend – from initial creation to maintenance efforts to phasing out
- The pilet's functional life cycle – from loading to installation to uninstallation

Let's try to cover both a bit here, with the focus being on the latter to fully understand how a pilet works.

We have already scaffolded the pilet (in the previous section). Once this has been completed, we can start the development of our desired functionality. To start the debugging process, we can use the Piral CLI to start a development server with live reloading as follows:

```
npx pilet debug
```

At some point, our micro frontend will be feature-complete for a first release. Now, we should first try to build it:

```
npx pilet build
```

Alternatively, if we have already a feed server running somewhere, we can also publish it directly. This works against a created npm package or by doing everything in one command:

```
npx pilet publish --fresh --api-key <your-api-key> --url https://feed.
piral.cloud/api/v1/pilet/<feed-name>
```

As a target feed, we will choose the same feed that was used as the source for all the pilets within our final app shell.

Updates to this micro frontend are developed and released with the same three commands. We first debug and fix the issue, then we try to build. If everything goes well, we can then publish the updated pilet. The only thing to make sure of is to change the version within the package.json file. Just like normal npm packages, we can publish every pilet just once per version. One of the reasons for this is to ensure that pilets can be cached indefinitely, which makes them really efficient frontend resources.

To be efficient, we should leave the actual publish steps to a CI/CD pipeline. Since we used GitHub for the app shell, we could also host this pilet in a GitHub repository. Therefore, using GitHub Actions is a straightforward way to establish a pipeline quickly.

The .github/workflows/publish.yml file of the pilet uses publish-pilet-action to leverage the Piral CLI. The API key for publishing is made available through GitHub secrets as follows:

```
name: CI

on:
  push:
    branches:
      - main

jobs:
  publish-pilet:
    name: Build and Deploy
    runs-on: [ubuntu-22.04]
```

```
    steps:
    - uses: actions/checkout@main
    - name: Publish Pilet
      uses: smapiot/publish-pilet-action@v2
      with:
        feed: <feed-name>
        api-key: ${{ secrets.apiKey }}
```

Now, let's say that the pilet was published successfully. What happens if a user accesses the app shell? Like before, the feed server is called when the loader script is initialized. In contrast, however, the response will not be empty but will instead contain the published pilet.

The response from the feed server may now look similar to this:

```
{
  "items": [
    {
      "name": "<pilet-name>",
      "version": "<pilet-version>",
      "link": "https://assets.piral.io/pilets/<feed>
        /<pilet-name>/<pilet-version>/index.js"
    }
  ]
}
```

This tells the app shell that there is a single micro frontend that should be loaded from the URL mentioned in the `link` property.

Once the script has been evaluated, the Piral framework will look for a special function that was exported from the script: `setup`. This function is needed for integration purposes. It will be called from Piral with the API object as the only argument.

Piral knows three phases for each micro frontend:

- Evaluation (the micro frontend is loaded)
- Setup (the micro frontend is integrated)
- Teardown (the micro frontend is removed)

While the first phase is implicit, the second phase is required for a script to be recognized as a micro frontend. If a script does not export a `setup` function, it cannot be integrated further. On the other hand, a `teardown` function is optional and not really needed.

One of the reasons why a `teardown` function may be useful is to clean up some resources that may cause memory leaks or conflicts if the micro frontend is removed or updated. This way, we can remove, e.g., event listeners or WebSocket connections that are bound to the lifetime of the micro frontend.

Now it's time to actually create and use components in our micro frontend solution.

Implementing framework-agnostic components

The `index.jsx` file of a pilet starts like this:

```
export function setup(api) {
  // ...
}
```

If we want to register a new page, we can use the `registerPage` function exposed by the API. This function expects two arguments: the path that triggers the navigation and the component to show as content. This is as follows:

```
export function setup(api) {
  api.registerPage('/example-page', PageComponent);
}
```

The component itself can be defined inline or referenced from some other module. Ideally, we want to use some kind of lazy loading to only load the code for the component when it should be rendered.

Lazy loading in micro frontends improves performance by loading components only when needed, reducing the initial load time and resource consumption. This approach enhances the user experience by delivering content more efficiently and enables better scalability as the application grows in complexity and size.

With React, such a lazy loading mechanism could be implemented like this:

```
import * as React from 'react';
const PageComponent = React.lazy(() => import('./Page'));
```

Since Piral has first-class support for React, there is nothing else to do in this case. It just works. However, what about components coming from other frameworks, such as Svelte for instance?

Let's pretend that we've already configured webpack to process Svelte modules and that we've already created `PageComponent` in Svelte:

```
import PageComponent from './Page.svelte';
```

Registering a component coming directly from Svelte is not possible. Instead, we need to wrap it using a generic life cycle:

```
export function setup(api) {
  api.registerPage('/example-page', {
    type: 'html',
    component: {
```

```
        mount(container, data) {
          container.$svelte = new PageComponent({
            target: container,
            props: {...data},
          });
        },
        update(container, data) {
          Object.keys(data).forEach((key) => {
            container.$svelte[key] = data[key];
          });
        },
        unmount(container) {
          container.$svelte.$destroy();
          container.innerHTML = '';
        },
      },
    });
}
```

This defines the whole DOM life cycle for the component. Since the code is quite lengthy, it may make sense to extract that into a dedicated function. Luckily, there is already a Piral plugin that does that. It can be also used directly from a pilet.

Installing `piral-svelte` as a dependency in the pilet makes the following code possible:

```
import { fromSvelte } from 'piral-svelte/convert';
import Page from './Page.svelte';

export function setup(api) {
  api.registerPage('/example-page', fromSvelte(Page));
}
```

Likewise, other frameworks can either be integrated simply by using an existing plugin or by building upon the generic HTML component with the explicit life cycle definition.

While siteless UIs provide a powerful model for micro frontend solutions that are composed on the client, the concepts behind it can also be brought to the server. Building on top of the central rendering server from *Chapter 7* and the optimization of progressive rendering introduced in *Chapter 10*, we can form a new pattern: the **islands composition**.

> **Islands architecture and progressive rendering**
>
> As the name islands composition already hints, it is based on the islands architecture, as mentioned at the end of *Chapter 7*. While islands architecture has a huge overlap with progressive rendering, it is not a descendant or specialization of it. In fact, a web app using islands architecture can also be implemented using a different pattern than progressive rendering. For instance, Qwik uses a pattern called **resumability** to bring islands architecture to the client efficiently. While the identification and deferred loading are certainly part of an islands architecture implementation, it also brings other features such as event-based loading and orchestration to the table.

In the following section, we'll see how islands composition can be implemented by using the popular server-side framework Qwik as a basis.

Implementing islands composition with Qwik

In order to implement islands composition, we can either start from scratch or on top of an existing framework. As there are many frameworks that enable islands architecture already, it makes sense to sketch an approach that starts with an existing monolithic approach, ultimately extending this with micro frontend capabilities.

One of the frameworks that starts as monolithic but has everything that it needs to be adjusted and used as a basis for a distributed web application is Qwik. Qwik is a modern web framework that is based on a novel approach referred to as resumability. Using resumability, the framework is capable of instantly making any part of the page interactive if needed. As developers, we mostly don't need to care which parts need to be made interactive or when. All these decisions are made by the compiler, which is capable of making bundle splitting on the functional level.

Extracting parts of individual functions yields a much finer granularity than splitting on a component level. This way, the framework is no longer bound to the component model or to developers' decisions on how the application is structured in terms of components.

For implementing islands composition with Qwik, we can choose between different implementation paths. In the direct approach, we might want to alter Qwik to be able to retrieve and handle remote components. While this approach would have many benefits, it also ends up with the most development effort. Therefore, it could be a viable approach for a larger team with a lot of experience using Qwik, but definitely not for mid-size to smaller applications with smaller teams or teams with no experience with Qwik.

Another development approach is to treat each micro frontend as a separate Qwik process, that is, an individual server. This goes all the way back to having separate services running as outlined in *Chapter 7*. The only difference is that the requested HTML will be enhanced with Qwik instructions, resulting in resumable snippets that follow islands architecture by itself.

The contrast between islands composition and the siteless UI pattern is best seen by looking at the following architecture diagram for islands composition:

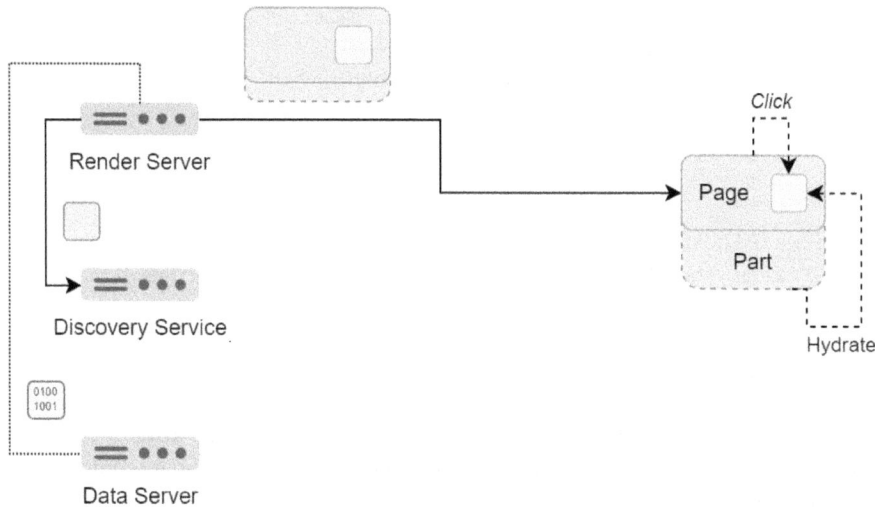

Figure 11.2 – Islands composition

In *Figure 11.2*, we can see that the most important difference compared to the siteless UI diagram outlined in *Figure 11.1* is the pre-rendering of the full page on a dedicated render server. There is no full orchestration script available, even though some loader script to manage the hydration (if needed) for components from the different micro frontends might be required.

Islands composition with siteless UIs

One question that might arise is as follows: is islands composition really standalone or is it an extension that can be added to most patterns? While there are certain arguments that most patterns can be extended to support the islands architectures pattern, it makes the most sense to view islands composition as a standalone pattern. The best argument for this view is that for most patterns, the original architecture diagram has to be altered significantly to support islands architecture. In the case of siteless UIs, this is certainly also the case. However, as we'll see in *Chapter 15*, the augmentation is not so significant, making applications using the siteless UI pattern quite easy to adopt and migrate.

To use a component from an available micro frontend, we need to reference it in a special component – just like in the siteless UI example with the `registerExtension` function. The main difference is that we've stayed framework agnostic beforehand, therefore using a rather generic function, while we now do everything in Qwik, effectively using Qwik's component model.

A wrapper component to provide this functionality could be defined as follows:

```
import { component$, SSRStream, SSRStreamBlock } from '@builder.io/
qwik';

export default component$(({ remote, removeLoader = false }) => {
  const url = remote.url;
  // ...

  return (
    <SSRStreamBlock>
      <SSRStream>{getStream(url, removeLoader)}</SSRStream>
    </SSRStreamBlock>
  );
});
```

Using the two provided `remote` and `removeLoader` props, a special kind of component is created – one that uses a streaming block for actually asynchronously creating its content. This way, performance is preserved as best as possible.

The special `SSRStream` component from Qwik expects a stream function to be present as its children. To obtain the stream function, we introduced a helper function named `getStream`.

The `getStream` function can be defined as follows. It returns a function that has a single argument representing a `StreamWriter`. The function must return a `Promise` to be valid:

```
function getStream(remoteUrl, removeLoader) {
  return async (stream) => {
    const remoteUrl = new URL(remoteUrl);

    if (removeLoader) {
      remoteUrl.searchParams.append('loader', 'false');
    }

    const res = await fetch(remoteUrl, {
      headers: { accept: 'text/html' },
    });
    const reader = res.body.getReader();
    const decoder = new TextDecoder();
    let fragmentChunk = await reader.read();
```

```
    let base = '';

    while (!fragmentChunk.done) {
      const rawHtml = decoder.decode(fragmentChunk.value);
      const fixedHtmlObj = fixRemoteHTMLInDevMode(rawHtml, base);
      base = fixedHtmlObj.base;
      stream.write(fixedHtmlObj.html);
      fragmentChunk = await reader.read();
    }
  };
}
```

The `removeLoader` parameter can be used to distinguish between the fragment case and the standalone scenario. In the former scenario, the Qwik loader is already present from the consuming component, which is most likely a page. Consequently, we don't need to add the special Qwik loader script to the generated HTML. In the other case, we will still need to append the loader script.

The rest of the code just deals with retrieval of the component's HTML – any modifications to the received HTML, as well as appending the modified HTML to the output. Finally, this will result in an HTML block that is fully resumable from Qwik, as it was enhanced with all the necessary information.

How can this block now be used in practice? Well, let's say we have a page that consists of two parts. One part is present on all pages in the header region, while the other part is in the main content canvas. The latter should be displayed by a micro frontend, while the former already comes from the app shell.

The code that follows has been set up to solve this scenario. As an example, we'll consider a shopping website that uses a micro frontend to show the list of products – just like the tractor store we used as an example in previous chapters such as *Chapter 7* – from a micro frontend. The shopping cart icon in the header bar is rendered from the app shell.

Let's see this in action:

```
import { $, component$, useOnDocument, useSignal } from '@builder.io/
qwik';
import { CartCounter, RemoteMfe } from 'shared/ui';
import { CartChangedEvent } from 'shared/constants';
import { remotes } from 'shared/remotes';

export default component$(() => {
  const cartQtySignal = useSignal(0);
  useOnDocument(CartChangedEvent, $((event) => {
    cartQtySignal.value += (event as CustomEvent).detail.qty;
  }));

  return (
```

```
  <>
    <div class="header-bar">
      <CartCounter count={cartQtySignal.value} />
    </div>
    <RemoteMfe remote={remotes.home} removeLoader />
  </>
  );
});
```

The code itself wires up a DOM event to change the quantity value of the cart. The other part of the event – the emitter – is not included in the code. Instead, it comes with the rendered micro frontend, that is, from a different server. This works, as the whole construction of the frontend is distributed. However, for the communication between the different fragments, the platform shared by all of them is used: the DOM.

With this in mind, let's wrap it up to continue our journey with some more practical tips and strategies.

The code for this section can be found at: `https://github.com/PacktPublishing/The-Art-of-Micro-Frontends-Second-Edition/tree/main/Chapter11/11-islands-composition`

Summary

In this chapter, you learned how to create appealing SPAs using individual plugins, which are merged into a single coherent web application. In contrast to the classic plugin architecture, these modules may get privileges and responsibilities that are on the application level.

As with the SPA composition before, you saw that user interaction is meant to be quite high. However, unlike with SPA composition, the user experience and developer experience are supposed to be smooth. While the internal complexity and tool reliance have increased, the major drawback is the dependency of the modules on the API defined in the app shell.

In the next chapter, we'll dive deeper into runtime integration of micro frontends. In particular, we want to explore how we can efficiently share dependencies at runtime using modern mechanisms such as webpack's Module Federation plugin.

Part 3:
Bee Brood – Implementation Details

In this part, you will gain in-depth knowledge for almost all remaining questions on actual micro frontend implementations. You will be able to efficiently share dependencies, isolate styling, secure your application, as well as distribute and aggregate micro frontends in a central service – all with modern standards and best practices in mind.

This part covers the following chapters:

- *Chapter 12, Sharing Dependencies with Module Federation*

- *Chapter 13, Isolating CSS*

- *Chapter 14, Securing the Application*

- *Chapter 15, Decoupling Using a Discovery Service*

12

Sharing Dependencies with Module Federation

In the previous chapter, we investigated the advantages and use cases for the siteless UI pattern. At its core, the siteless UI pattern is made specifically for large-scale client-side composed applications. Naturally, one of the most frequently occurring challenges with any kind of client-side composed micro frontend solution is third-party dependencies.

In this chapter, we'll discuss what makes third-party dependencies special. We'll see that some tools and methods exist to deal with third-party dependencies efficiently – at compile time and at runtime. Ultimately, the goal of this chapter is to teach you to be able to use these tools for crafting the best possible micro frontend solutions.

We'll cover the following topics in this chapter:

- Sharing dependencies between micro frontends
- Utilizing Module Federation
- Understanding Native Federation
- Achieving independence with SystemJS

The techniques presented in this chapter will help us build systems that not only scale organizationally but also regarding performance.

Technical requirements

To follow the code samples in this book, you need to have knowledge of JavaScript and how to use the command line. You should have Node.js (version 20 or higher) installed using the instructions at `https://nodejs.org`. For this chapter, additional knowledge in using the webpack bundler can be helpful.

The code used in this chapter can be found in the following repo:

`https://github.com/PacktPublishing/The-Art-of-Micro-Frontends-Second-Edition/tree/main/Chapter12`

The CiA video for this chapter can be found at `https://packt.link/pIaGB`

Sharing dependencies between micro frontends

One of the most important considerations when dealing with micro frontends is where they are composed. We've seen in many of the previous chapters that the composition can happen on the server (*Chapter 7*) or the client (*Chapter 9*). Generally, however, no matter where we decide to actually stitch the different fragments together, we'll bring these parts to the browser.

In many cases, we don't only deliver HTML and CSS to the browser but we also bring over some JavaScript. This is where things start to get interesting. While there are certain scenarios wherein micro frontends are either restricted to bringing JavaScript or what JavaScript they can deliver, we will generally have to deal with pretty much anything in there. This includes some third-party libraries such as React or jQuery.

Like CSS, the major problem with JavaScript is that it is globally shared by default. Unlike CSS, with shadow DOM there is no good solution to properly isolate JavaScript. If we don't want to put UI fragments in a dedicated iframe, we will have to run all JavaScript in the same global DOM. This presents a challenge.

Let's consider the case of using jQuery as a third-party dependency. By default, jQuery uses the global $ variable as an identifier. This means that two micro frontends that independently use jQuery might run into a problem. There are cases where this problem is not really striking, such as when both versions are compatible and have no plugins, modifications, or other extensions for jQuery. However, in general, we'll always face the threat of running into trouble. So, what can we do here?

It turns out that there are various ways to circumvent the issue. One is to use a way that is specific to the third-party dependency in question. For jQuery, we could either use a modular version, which accesses the functionality via an `import` statement, or we could use a dedicated name for jQuery per micro frontend by utilizing jQuery's `noConflict` function.

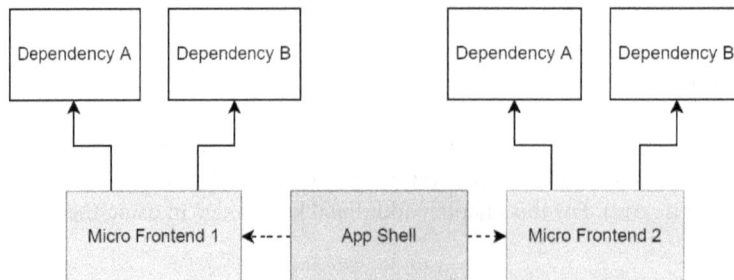

Figure 12.1 – No dependency sharing

In *Figure 12.1*, we can see that no sharing leads to each micro frontend coming up with its own version of a certain dependency. In the worst-case scenario, this means that code is unnecessarily duplicated and that runtime conflicts might arise. To avoid this, methods to efficiently share dependencies have to be introduced.

Another way is to realize that micro frontends might actually need to use jQuery. With this knowledge, the app shell could already be equipped with jQuery, making it a shared dependency. The problem with this approach is that the app shell should not have or need this kind of knowledge. We've transferred part of our micro frontend technicalities to the app shell, making the app shell's team responsible for maintaining the dependency. This is a difficult job. Therefore, this kind of sharing, referred to as **central dependency sharing**, has a few downsides that we want to get rid of.

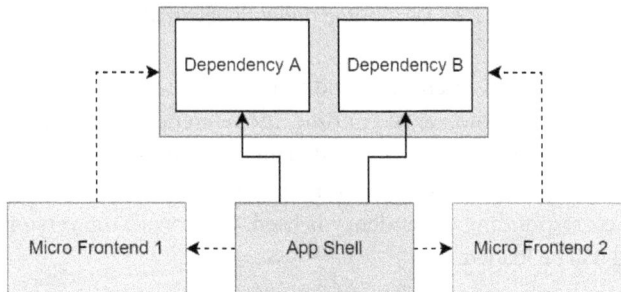

Figure 12.2 – Central dependency sharing

In *Figure 12.2*, we see that dependencies can be introduced and shared by the app shell already. This central sharing offers a lot of benefits compared to the share-nothing approach shown in *Figure 12.1*. Nevertheless, as the number of micro frontends and dependencies grows, this approach becomes less and less scalable.

A better option is to use **distributed sharing**. Distributed sharing works by delivering the necessary third-party dependencies along the micro frontend that needs it. The magic ingredient for distributed sharing is that before loading a dependency, a check is performed to see whether the same or similar dependency is already loaded. This check could allow for having multiple versions of the same dependency or only a single one. Likewise, it could allow for using a different version if, for example, the major version is the same.

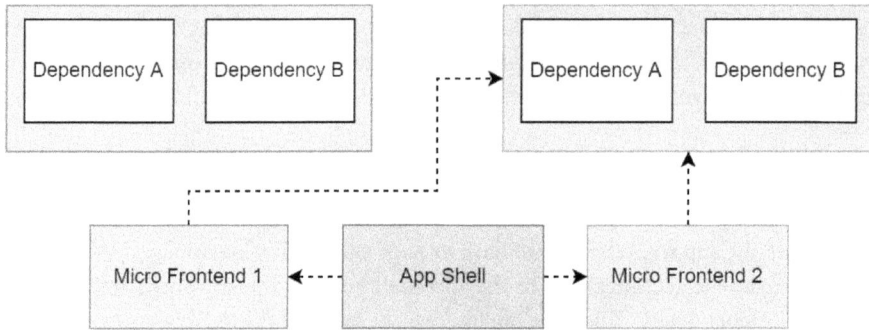

Figure 12.3 – Distributed dependency sharing

In *Figure 12.3*, we can see that each micro frontend might come with its own set of dependencies, which are decided at build time. However, at runtime, the references could be drawn in a way that improves performance and avoids conflicts. When dependencies need to be loaded, a special loader might be used to identify whether a compatible version has been loaded already. In the case of an evaluated module, the corresponding dependency is used. Otherwise, the version coming with the current micro frontend will be loaded.

One way to introduce distributed dependency sharing is by utilizing Module Federation tooling. Let's see how this looks in practice.

Utilizing Module Federation

Module Federation has been introduced as an integrated plugin for the webpack bundler. While Module Federation can be thought of as a mental model for micro frontends or dependency sharing, it is also, most importantly, a technical implementation and tooling to support micro frontends.

Classically, Module Federation has been part of the webpack bundler, starting from version 5. As of today, there are other bundlers such as **rspack** that support Module Federation. While rspack is still very close to webpack, the whole approach has been ported to API-incompatible bundlers too. Most notably, there is a community contribution that brings some of the features of Module Federation to the **Vite** bundler.

To get started with Module Federation, we need to use a supporting bundler. For example, let's say that we'll stay at the original implementation using webpack. To actually use webpack, we must create a valid webpack configuration, which lives in a file called webpack.config.js. This file can be minimalistic, as shown here:

```
const { resolve } = require('path');

module.exports = {
```

```
    mode: 'production',
    entry: './src/index.js',
    output: {
      path: resolve(__dirname, 'dist'),
      filename: 'app.js',
    },
};
```

The example configuration takes a JavaScript file found in src/index.js and follows all the references, such as module imports, of this file. The resulting code is then processed and written to dist/app.js. The resolve function from the Node.js path module is used to convert relative paths to absolute paths.

For Module Federation to be enabled, we need to extend the webpack configuration with a special plugin called ModuleFederationPlugin. This is a class that is defined on the container object exported directly from the webpack module. The container object is actually a collection of plugins and functions that are all related to the webpack Module Federation – just with different levels of abstraction.

If the application is essentially a micro frontend, then the previous webpack configuration can be changed as follows:

```
const { resolve } = require('path');
const { container } = require('webpack');
const { ModuleFederationPlugin } = container;

module.exports = {
  mode: 'production',
  entry: {
    remote: './src/index.js',
  },
  output: {
    path: resolve(__dirname, 'dist'),
    filename: 'app.js',
  },
  plugins: [
    new ModuleFederationPlugin({
      name: 'remote',
    }),
  ],
};
```

With these changes, the application makes an opt-in to Module Federation. In the world of Module Federation, this means that the application can share dependencies with other federated modules. In the provided case, the application could be a **host**, that is, it could load other modules. If we'd only changed the configuration – not the actual application code – we'd certainly not leverage Module Federation yet.

The other perspective is to also be a **remote**, which simply implies that we will also expose modules from our code base to be loaded by other Module Federation hosts. For this, our configuration could be altered to look like this:

```
// ... all as before
    new ModuleFederationPlugin({
      name: 'remote',
      exposes: {
        './app': './src/index.js'
      },
    }),
  ],
};
```

Now, any Module Federation host is able to load the bundled `./src/index.js` module by referring to `remote/app`. However, how could the webpack configuration of another micro frontend look to actually achieve this?

The `remotes` configuration specifies the remote modules that the host application can directly use. Each remote module is identified by a name and a URL pointing to its entry point. This setup allows the host to dynamically load and use modules from the remote, facilitating modular and scalable micro frontend architectures:

```
const { resolve } = require('path');
const { container } = require('webpack');
const { ModuleFederationPlugin } = container;

module.exports = {
  mode: 'production',
  entry: {
    host: './src/index.js',
  },
  output: {
    path: resolve(__dirname, 'dist'),
    filename: 'app.js',
  },
  plugins: [
    new ModuleFederationPlugin({
      name: 'host',
```

```
    remotes: {
      'remote': 'remote@https://url.com/remoteEntry.js'
    },
  }),
 ],
};
```

The configuration is very similar to the remote except that we now have a `remotes` section, which defines the remote modules that will be used in the codebase that is to be processed by webpack. In this case, we can just use the `remote/app` remote module by importing it.

The prefix (`remote`) has been chosen by us but usually matches the name of the remote, as chosen by the team behind it. The module's name (`app`) must match the name of the module chosen by its team. Most identifiers form some kind of contract and should not be renamed without prior alignment between the teams.

One important thing to note is that the provided URL has to match the location of the remote's entry point. The URL is dependent on the target server when the generated artifacts are deployed. The name of the file has to match the output of the created Module Federation entry point. While this can be named in the configuration provided to the `ModuleFederationPlugin`'s constructor, it makes sense to come up with some useful convention or to stick to the default value, which is `remoteEntry.js`.

Internally, an entry module created by Module Federation will expose two functions: `init` and `get`. The former is used to initialize the sharing and connect to the whole application. The `get` function is then used to obtain an exposed module using the path identifiers that have been configured for the actual build, such as `./app`. This entry module is called a container.

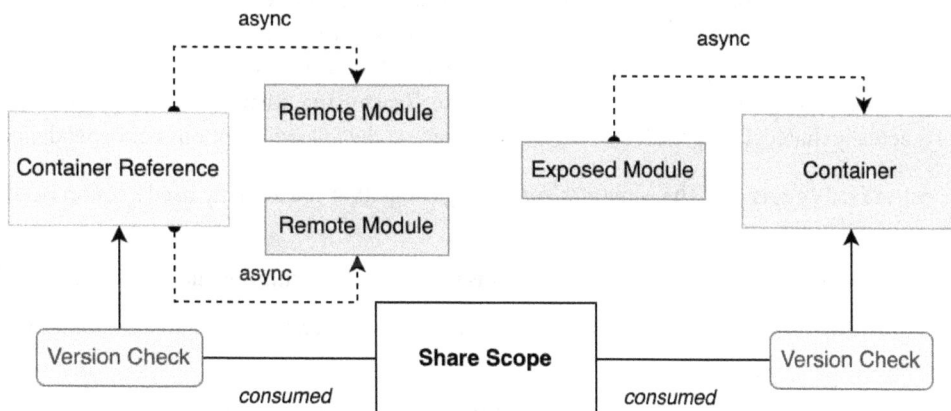

Figure 12.4 – The relationship between containers, container references, and the share scope

The relationship between containers, container references, and the share scope is illustrated in *Figure 12.4*. The general idea is that everything is connected via a common share scope, which is named `default` by default but could also be named differently, essentially allowing for the isolation of different modules, too.

A container reference is the reference to a remote, while the exposed modules are already available within the share scope. In both cases, version checks are implemented to properly assign and load dependencies.

Speaking of dependencies, the most important aspect of Module Federation is actually the capability for sharing dependencies. This is achieved via a special configuration key called `shared`:

```
new ModuleFederationPlugin({
  name: 'host',
  remotes: {
    'remote': 'remote@https://url.com/remoteEntry.js'
  },
  // example shared configuration to add "react" to the
  // list of shared dependencies
  shared: {
    react: {
      requiredVersion: '16.x',
      singleton: true,
    },
  },
}),
```

There are a few ways in which the shared option can be set, for example, as an array of strings. In the preceding example, we used the explicit form, which allows us to use the so-called sharing hints, that is, the options for defining how a dependency is actually shared and used. This form gives us the most control over how a certain dependency (in this case, React using the name of its npm package, `react`) is actually shared. Using these sharing hints, we can set the following options per dependency:

- `requiredVersion`: The semantic versioning range that the actually used version needs to fulfill
- `version`: The version of a dependency that is being brought with the micro frontend
- `singleton`: This is `true` if only a single instance of the dependency should exist, otherwise it is `false`
- `strictVersion`: This is `true` if an exception should be thrown when another version of a singleton should, but cannot be, loaded; otherwise, it is `false`
- `eager`: This is `true` if the dependency should always be available and `false` if it should be loaded asynchronously

- `import`: Specifies the fallback or local module to load (or import) when the dependency is required; set to `false` to avoid bringing the dependency with the micro frontend

- `packageName`: The actual name of the package

- `shareKey`: The name of the module in the share scope

- `shareScope`: The name of the share scope to use (standard is `default`)

While the default setting of `eager` has some impact on how the application is structured, it usually makes sense to keep the flow of modules asynchronous. This way, duplicates are avoided and everything runs smoothly.

A popular combination of sharing hints is to set `singleton` to `true` and `strictVersion` to `false`. Most of the time, one reason to use dependency sharing is to avoid conflicts, as mentioned. By enabling both settings, we'd have a single instance of the dependency, but we'd throw exceptions for micro frontends demanding non-fitting versions of the dependency.

Disabling the strict version check will still place a warning in the console with the available dependency version being taken. The warning is relevant as the taken version does not match the required version, but as the dependency was specified as a singleton, this is the only way to use it.

Many of the other options only make sense for edge cases. For instance, providing no fallback module or changing the package name does not make sense for standard operations. Even the `version` option only comes in handy if we don't have a proper npm module as a dependency. Most of the time, a `package.json` file for the module exists, which allows Module Federation to automatically read out and use the listed version.

Tight integration with webpack can limit flexibility and increase complexity, especially for projects that may benefit from different or no bundlers. Native Federation offers a solution by decoupling the dependency-sharing mechanism from specific bundlers, allowing greater flexibility and broader adoption of the host-remote model across various toolchains.

Understanding Native Federation

A common misconception is that Native Federation is compatible with Module Federation. Unfortunately, these technologies are incompatible with each other and would require an orchestrator library such as **Picard** to work together. Nevertheless, what remains the same is the mental model. Just like in Module Federation, the host-remote philosophy is also present in Native Federation.

The Native Federation package only contains lower-level functions. This allows framework authors to use these building blocks for actually constructing useful abstractions on top of it. One example is the way to actually load a shared module. Since there is no direct bundler integration, no direct imports are possible. Through the `loadRemoteModule` function, we can retrieve the module using the same notation as Module Federation, that is, by referring to the name of a remote and the path to the requested exposed module.

Let's look at an example code for loading and using a remote module:

```
import { loadRemoteModule } from '@softarc/native-federation';

loadRemoteModule({
  remoteName: 'red',
  exposedModule: './productPage',
}).then(({ renderProductPage }) => {
  const root = document.querySelector('#app');
  renderProductPage(root);
});
```

In the code, we import the `loadRemoteModule` helper from the Native Federation npm package. We use the helper to start loading the exposed product page module of the `red` remote.

Importantly, once the promise resolves with the results of loading the requested module, we can do something such as rendering the obtained component or calling the obtained function. In case of a network issue or an invalid path, we'd see a rejected promise. So, keep in mind to also take care of the error handling.

For actually initializing Native Federation, we need to use the `initFederation` function. This function expects either an object containing a hard-wired mapping to the remotes to load, or a string representing a URL leading to a manifest file. The manifest file will be in the same format as the object containing the hard-wired mapping.

In short, the initialization with an object might look as follows:

```
import { initFederation } from '@softarc/native-federation';

await initFederation({
  'red': 'http://localhost:2001/remoteEntry.json',
  'blue': 'http://localhost:2002/remoteEntry.json',
  'green': 'http://localhost:2003/remoteEntry.json',
});
```

This code tells Native Federation where the individual modules live. Consequently, Native Federation starts loading the JSON files using the information received to the `initFederation` function. The JSON files then serve as entry points for the individual micro frontends. In contrast to Module Federation, the entry point only contains information about the shared dependencies, not about other remotes that are being used. Therefore, the whole JSON is incomplete at first and will only be completed at runtime.

The most crucial difference between Module Federation and Native Federation is that Native Federation is not wrapped around webpack's module format. Instead, the basic idea behind Native Federation is to use the native module format of JavaScript, which is called **ES Modules** (**ESM**). For ESM to use dependency sharing, a novel mechanism called import maps is required. While third-party scripts exist to bring this mechanism to unsupported browsers, the best possible outcome requires direct browser support.

Like Module Federation, the actual dependency-sharing part of Native Federation has to be done via the build process. As Native Federation is bundler-independent, the way to configure this might differ between the various bundlers. For the official esbuild adapter of Native Federation, dependency sharing can be achieved using a configuration file. Usually, the configuration file is named `federation.config.js`, but this could be changed, too:

```
const {
  withNativeFederation,
  shareAll,
} = require('@softarc/native-federation/build');

module.exports = withNativeFederation({
  name: 'remote',
  exposes: {
    './module': './src/index.ts',
  },
  shared: {
    ...shareAll({
      singleton: true,
      strictVersion: true,
      requiredVersion: "auto",
      includeSecondaries: false,
    }),
  },
});
```

The `shareAll` helper function just takes all dependencies and adds them to the build process as shared dependencies. The provided object is used as a configuration for the found dependencies. In the case of our preceding example, we treat all dependencies as singletons with a strict version check applying the automatically discovered version as a requirement for the check. The transitive dependencies are not shared directly.

Native Federation's reliance on ESM and import maps can limit dynamic loading capabilities and compatibility with older browsers. SystemJS addresses these limitations by providing a flexible module loader that supports various module formats and dynamic loading, allowing for greater adaptability and ease of use in different environments.

Achieving independence with SystemJS

SystemJS is a pluggable module loader that is capable of dealing with a variety of formats. It describes itself as a dynamic ES module loader with extra functionality such as **Asynchronous Module Definition (AMD)** and import map support, or hooks to introduce, such as a custom resolution mechanism.

Unlike the previously discussed solutions (Module and Native Federation), SystemJS is almost entirely a runtime construct. However, since the runtime construct relies on a certain structure in the code, we might still want to use some tooling for obtaining valid SystemJS modules. For instance, using the webpack bundler, we need to change the output format to system. Generally, we do not need a bundler at all and could write valid SystemJS modules ourselves, too.

For example, take the following code, which will not be handled by any bundler and could be used directly:

```
System.register(['react'], function (_export, _context) {
  let react;
  return {
    setters: [ (_react) => {
      react = _react;
    }],
    execute: () => {
      const MyComponent = () => react.createElement('div', {}, 'Hello
World!');
      _export({
        MyComponent,
      });
    }
  };
}
```

The format of a SystemJS module has been tailored to allow asynchronous loading of dependencies. This is ideal for sharing dependencies in micro frontends. Also, as SystemJS is a pure runtime loader, it can be fully extended to go beyond its original use cases. Right now, SystemJS can be used on the server, on the client, and specifically for sharing dependencies in micro frontends.

As an example, consider having the previously outlined SystemJS module rewritten to look like the following:

```
System.register(['react@16.x'], function (_export, _context) {
  // content as beforehand
}
```

The content of the actual module did not change. However, the resolution in SystemJS is now using the react@16.x identifier instead of just react. The standard resolution of SystemJS just looks at all registered modules – in case of a direct hit, the module is either loaded or returned directly.

In the case of `react@16.x`, we'd need to split the identifier into two parts, resulting in a name (`react`) and version range (`16.x`). Now we'd need to compare this to the registered dependencies to identify where to get a compatible module from. Finally, we'd change the original identifier to match the corresponding registered module.

In code, this change can be added to SystemJS like this:

```
const systemResolve = System.constructor.prototype.resolve;

System.constructor.prototype.resolve = function (id, parentUrl) {
  try {
    // call the original - let's see if we can resolve it
    return systemResolve.call(this, id, parentUrl);
  } catch (ex) {
    // let's check if we can just find a match
    const result = findMatchingPackage(id);

    // does not look so - let's re-throw
    if (!result) {
      throw ex;
    }

    return result;
  }
};
```

Granted, much of the logic is located in the `findMatchingPackage` function. However, in general, the idea of redirecting the resolution to provide dynamic support is sound and very helpful. If all micro frontends are based on SystemJS, then this works quite nicely.

One framework that follows this approach is Piral. In Piral, every micro frontend compiles to SystemJS with proper support for distributed dependencies. Consequently, distributed dependencies work in the browser and in Node.js. To effectively do this, Piral identifies shared dependencies during the build process and renames the default identifiers (such as `react`) in the generated SystemJS modules with their combined `name@version` counterparts.

As with the other approaches for sharing dependencies, there are pros and cons to using SystemJS. In general, what you should look for is a mechanism that allows for sharing without having strong constraints that cannot be easily circumvented. Otherwise, you might end up in a situation where you either cannot establish some desired functionality or need to have a massive refactoring involving all your micro frontends.

Summary

In this chapter, you learned how you can leverage different libraries and tools to efficiently share dependencies between your micro frontends at runtime in the client. Most of these approaches are based on some assumptions, for example, that a certain module format has been taken.

In the next chapter, we will consider another topic that becomes quite important at runtime in the browser: how styles can be defined such that they don't collide with each other. This is especially important to avoid issues that only appear when certain combinations of micro frontends occur, which would make the whole solution unpredictable and not really able to scale well.

13
Isolating CSS

In the previous chapter, we identified the various kinds of third-party dependencies that we might encounter in micro frontends. In most cases, these dependencies only consist of scripts, but in some cases, they might also come with their own stylesheets. Especially in the case of component libraries, these styles might be wanted and should be shared among all used components. Nevertheless, the case that either a third-party dependency or a micro frontend ends up introducing some unwanted styling that impacts the whole layout is very likely.

One of the problems with micro frontend composition outside of the pattern introduced in *Chapter 6* is that all micro frontends are running in one document, which implies the sharing of all introduced stylesheets.

In this chapter, we'll look at methods to avoid running into style conflicts. We'll see how we can craft robust solutions that are flexible yet isolated and therefore immune to leaking styles. An important cornerstone of this goal is the usage of modern web standards and the capabilities of CSS.

We'll structure this chapter as follows:

- Understanding the consequences of open styling
- Implementation techniques to scope CSS
- Using the shadow DOM
- Using modern CSS features for isolation

Using the knowledge from this chapter, we'll be able to build solutions that will not suffer from leaking CSS stylesheets.

Technical requirements

To follow the code samples in this book, you need knowledge of JavaScript and how to use the command line. You should have Node.js (version 20 or higher) installed using the instructions at https://nodejs.org. For this chapter, additional knowledge of modern CSS can be helpful.

The code used in this chapter can be found at the following repo:

`https://github.com/PacktPublishing/The-Art-of-Micro-Frontends-Second-Edition/tree/main/Chapter13`

The CiA video for this chapter can be found at `https://packt.link/b2sLw`

Understanding the consequences of open styling

Putting components from different teams together on the same screen, i.e., within the same DOM, can result in the same problems that we'd face if we just append any styling to a monolithic solution. Conflicts may arise and crucial styles may be altered or removed. Consequently, we'll require a solution to prevent teams from just changing these styles.

Well, the first – and maybe most obvious – solution is to not have any special treatment, that is, we just leave styling completely open to all teams such that everyone can just add any styles to the shared document. If we follow this approach, each micro frontend can come with additional stylesheets that are then attached when the components from the micro frontend are rendered.

Open styling can lead to specificity conflicts and global style overrides, making it essential to implement scoped CSS solutions. This ensures that styles remain contained within their respective micro frontends. Leaving styling completely open can also lead to significant style collisions and maintenance challenges.

Ideally, each component only loads the required styles upon first rendering. However, since any of these styles might conflict with existing styles, we can also pretend that all problematic styles are loaded when any component of a micro frontend renders.

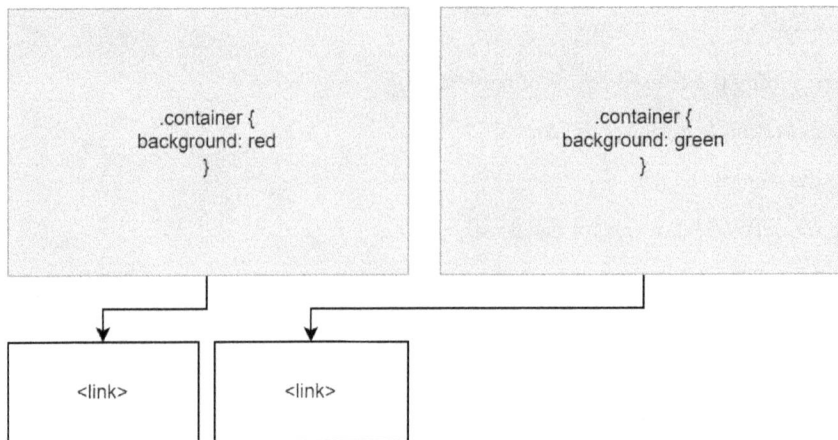

Figure 13.1 – No consideration of styling isolation will lead to styling collisions

The problem with this approach is shown in *Figure 13.1*. By using generic selectors such as `div` or `div a`, we'll potentially also restyle other elements, not just the fragments of the originating micro frontend. Even worse, using CSS classes and attribute selectors are no failsafe guard either depending on the used naming. A CSS class like `.container` is pretty generic and might therefore also be used in another micro frontend.

To prevent such issues, we need to apply some special implementation techniques for isolating CSS, i.e., put the styling into a scope dedicated to the micro frontend where the styles originate. Let's see how this could be done.

Implementation techniques to scope CSS

If every micro frontend follows a global CSS convention, then conflicts can be avoided on the meta level already. The easiest convention is to prefix each class with the name of the micro frontend. So, for instance, if one micro frontend is called *shopping* and another one is called *checkout* then both would rename their active class to `shopping-active` / `checkout-active` respectively.

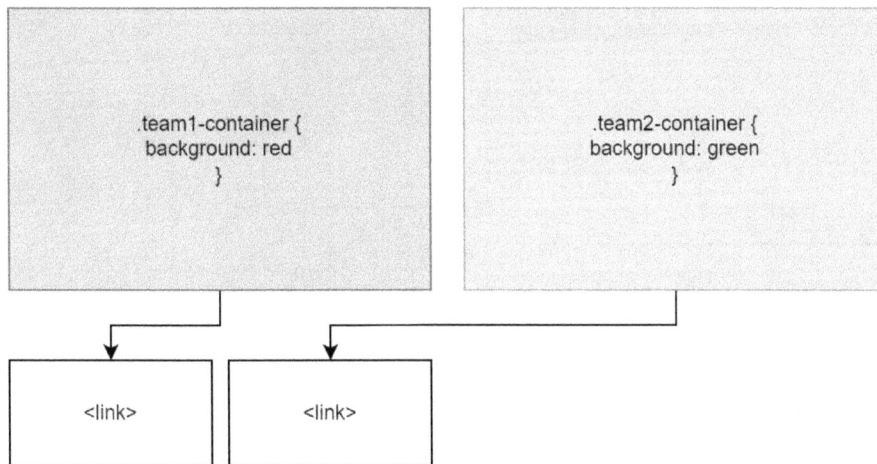

Figure 13.2 – Prefixing CSS class names by convention to avoid collisions

The same can be applied to other potentially conflicting names, too. As an example, instead of having an ID such as `primary-button`, we'd call it `shopping-primary-button` in the case of a micro frontend called *shopping*. This is also shown in *Figure 13.2*.

If, for some reason, we need to style an element such as a `div` or `img`, we should use descendent selectors such as `.shopping img` to style the `img` tag of a micro frontend called *shopping*. This now applies to `img` elements within any ancestor element having the `shopping` class. The problem with this approach is that the *shopping* micro frontend might also use elements from other micro frontends.

How would the styling look for a CSS selector such as `.shopping .checkout img`? Even though `img` is now hosted or integrated by the component brought through the *checkout* micro frontend, it would be styled by the *shopping* micro frontend CSS. This is not ideal as the original styles should be preserved.

While naming conventions solve the problem to some degree, they are still prone to errors and cumbersome to use. What if we rename the micro frontend? What if the micro frontend gets a different name in different applications? What if we forget to apply the naming convention at some points? This is where tooling helps us.

One of the easiest ways to automatically introduce some prefixes and avoid naming conflicts is to use CSS modules. Depending on your choice of bundler, this is either possible out of the box or via some config change.

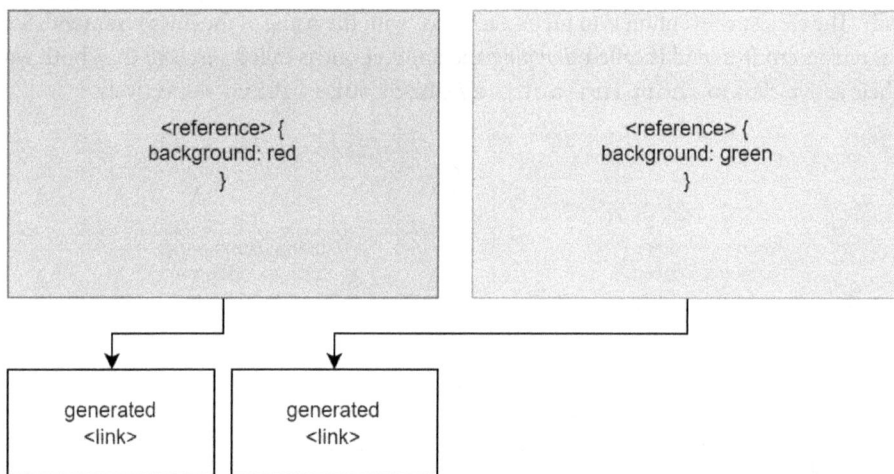

Figure 13.3 – CSS modules introduce generated class names to prevent collisions

In *Figure 13.3*, we see that the imported module is a generated module holding values that map to their original class names (e.g., `active`) to generated class names. The generated class name is usually a hash of the CSS rule content mixed with the original value. This way, the name should be as unique as possible.

In code, CSS modules might be used as follows:

```
// Import "default export" from CSS
import styles from './style.modules.css';

// Apply, e.g., in React
<div className={styles.active}>Active</div>
```

There are certain disadvantages that come with CSS modules. First, it comes with a couple of syntax extensions to standard CSS. This is necessary to distinguish between styles that we want to import (and therefore pre-process/hash) and styles that should remain as-is (i.e., to be consumed later on without any import). Another way is to bring the CSS directly into the JS files using a technique called **CSS-in-JS**.

CSS-in-JS has quite a bad reputation of late; however, I think this reputation is a bit of a misconception. Generally, we could also call it *CSS-in-Components* because it brings the styling to the components itself. Some frameworks such as Astro or Svelte even allow this directly via some other means, such as a `style` tag in the component.

The often-cited disadvantage of CSS-in-JS is performance, which is usually reasoned by composing the CSS in the browser. This, however, is not always necessary and in the best case the CSS-in-JS library is actually build-time-driven, i.e., without any performance drawback.

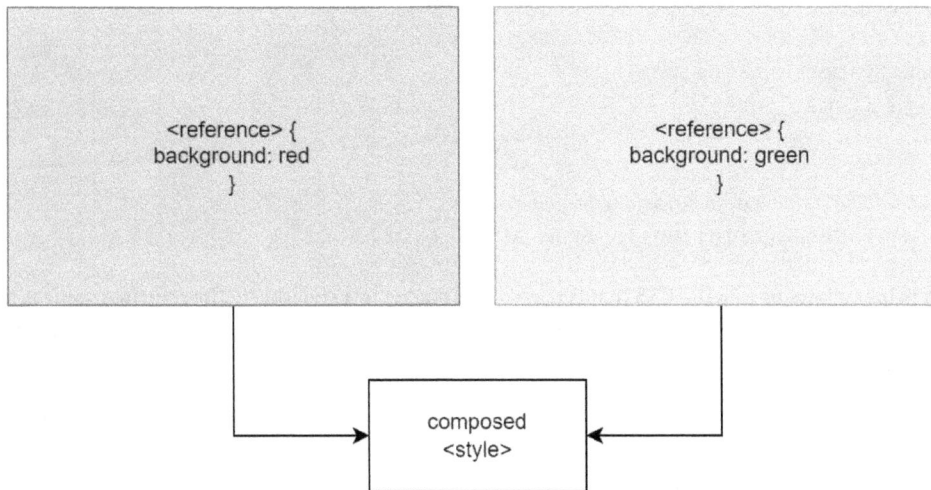

Figure 13.4 – CSS-in-JS generates collision-free class names for the specified declarations

In *Figure 13.4*, we see that CSS-in-JS is actually quite similar to CSS modules. However, the main idea is that no intermediate module is necessary to transport the knowledge about the generated CSS classes. Instead, everything is referenced directly.

When we talk about CSS-in-JS (or CSS-in-Components for that matter) we also need to consider the various options out there. For simplicity, I've only included the popular **Emotion** library. There are many others, such as **StyleX**, **Styled Components**, and **Vanilla Extract**. Let's see how these libraries can help us avoid conflicts when bringing together micro frontends in one application.

Emotion is a very cool library that comes with helpers for frameworks such as React, but without setting these frameworks as prerequisites. Emotion can be very nicely optimized and pre-computed and allows us to use the full arsenal of available CSS techniques.

Using *pure* Emotion is rather easy; first we need to install the package:

```
npm i @emotion/css
```

Now we can use it in our code as follows:

```
import { css } from '@emotion/css';

const tile = css`
  background: blue;
  color: yellow;
  flex: 1;
  display: flex;
  justify-content: center;
  align-items: center;
`;


// later, e.g., in a React component
<div className={tile}>Hello from Blue!</div>
```

The css helper allows us to write CSS that is parsed and placed in a stylesheet. The returned value is the name of the generated class. Since the helper generates the CSS on the fly, it will properly encapsulate it, making it pretty much impossible to override without being explicit about it.

If we want to work with React in particular, we can also use the jsx factory from Emotion (introducing a new standard prop called css) or the styled helper:

```
npm i @emotion/react @emotion/styled
```

This now feels a lot like styling is part of React itself. For instance, the styled helper allows us to define new components:

```
import styled from '@emotion/styled';

const Output = styled.output`
  border: 1px dashed red;
  padding: 1rem;
  font-weight: bold;
`;


// later
<Output>I am groot (from red)</Output>
```

In contrast, the `css` helper prop gives us the ability to shorten the notation a bit:

```
<div css={`
  background: red;
  color: white;
  flex: 1;
  display: flex;
  justify-content: center;
  align-items: center;
`}>
  Hello from Red!
</div>
```

All in all, this generates class names that will not conflict and will provide the robustness of avoiding a mix-up of styles. The `styled` helper in particular was inspired heavily by the popular Styled Components library.

Two other methods that you might find interesting are to use a CSS utility library such as Tailwind or a framework that supports isolated components – such as **Angular**, **Blazor**, or **Vue**. However, before resorting to some tooling, framework, or library, it might be useful to look for browser-native ways to solve the issue. One good way to isolate components is to use the shadow DOM that comes with the web components specification.

Using the shadow DOM

In a custom element, we can open a shadow root to attach elements to a dedicated mini document, which is actually shielded from its parent document. Overall, this sounds like a great idea, but like all the other solutions presented beforehand in this chapter, there is no hard requirement.

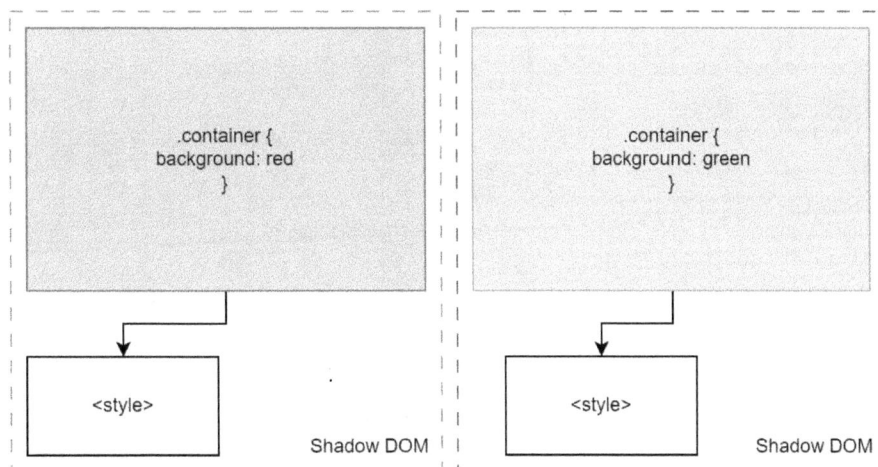

Figure 13.5 – Shadow DOM creates a mini document that is collision-free by default

In *Figure 13.5*, we see that the mini document of a shadow DOM isolates the contained DOM nodes from any outside styling. Therefore, by bringing in only the relevant styles from the shown micro frontend, we'd get automatic isolation. No collisions can happen inside the shadow DOM.

Ideally, a micro frontend is free to decide how to implement the components. Therefore, the actual shadow DOM integration has to be done by the micro frontend.

There are some downsides to using the shadow DOM. Most importantly, while the styles inside the shadow DOM stay inside, global styles also do not impact the shadow DOM. This seems like an advantage at first, but since the main goal of this whole chapter is to only isolate the styles of a micro frontend, we might miss requirements such as applying some global design system (e.g., using the **Bootstrap** library).

To use the shadow DOM for styling, we can either put the styles in the shadow DOM via a `link` reference or a `style` tag. Since the shadow DOM is initially not styled and no styles from the outside propagate into it, we'll actually need to do that – attach the corresponding tags for bringing in other styles. Besides writing some inline style, we can also use the bundler to treat `.css` (or maybe something like `.shadow.css`) as raw text. This way, we'll get just some text.

Now, styling something in the shadow DOM works like the following example:

```
import css from './style.css';

// define the custom element (web component)
customElements.define(name, class extends HTMLElement {
  constructor() {
    super();
    // once constructed we attach an open shadow DOM root
    this.attachShadow({ mode: 'open' });
  }

  // this method is called when the element is attached
  connectedCallback() {
    // we set the inline CSS to use display: contents
    this.style.display = 'contents';
    // create the style element and set its content
    const style = this.shadowRoot.appendChild(document.
createElement('style'));
    style.textContent = css;
  }
});
```

The preceding code is a valid custom element that will be transparent from the styling perspective using the `display: contents` mode, i.e., only its contents will be reflected in the render tree. It hosts a shadow DOM that contains a single `style` element. The content of the `style` is set to the text of the `style.css` file.

There are good reasons for avoiding shadow DOM for domain components. One reason is that not every UI framework is capable of handling elements within the shadow DOM. Therefore, an alternative solution has to be looked for anyway. One way is to fall back to using modern CSS features instead.

Using modern CSS features for isolation

In recent years, CSS has made huge leaps in improvement with respect to various areas including styling isolation. To achieve this, a couple of new constructs have been added to the CSS syntax:

- `@layer` gives us the possibility to define the precedence of CSS rules without relying on the initial ordering

- `@scope` makes it possible to define a new scope based on some selector

- `@container` is very handy for creating size-based rules based on an element

- `::slotted` allows us to style elements used in a shadow DOM differently; it must be declared inside the shadow DOM

- `::part` can be used to style any element inside the shadow DOM having a matching `part` attribute; it must be declared outside the shadow DOM

Naturally, any enhancement for the shadow DOM is an actual improvement for the support of micro frontends. However, even the constructs that are not directly related to shadow DOM may come in handy. Take, for instance, the possibility of influencing the CSS layering order. By using `@layer`, we finally have influence over the cascade, pretty much ruling out unclear specificity rules that may change at any time in the code without us noticing.

Let's see an example of where `@layer` might come in handy. This scenario is what we might have already faced in a stylesheet:

```
.overly#powerful .framework.widget {
  color: maroon;
}


.my-single_class { /* add some IDs to this ??? */
  color: rebeccapurple; /* add !important ??? */
}
```

If we want to override the color of an element that has multiple classes (e.g., `framework`, `widget`, and `my-single-class`) then simply having a primitive selector as shown in the preceding snippet might not work. In this case, another selector uses multiple classes and even an ID selector. The latter already provides a higher specificity than any class selector – but even without this, we are in trouble. After all, it's multiple classes versus a single one.

Quite often the only way out is to use the !important flag. But this way we only reversed the problem, making the opposite now more difficult than beforehand. This is the moment in time when @layer comes in to save the day:

```
@layer framework {
  .overly#powerful .framework.widget {
    color: maroon;
  }
}

@layer site {
  .my-single_class {
    color: rebeccapurple;
  }
}
```

Layers are great to bring in some additional structure to CSS. Most importantly, they give us the option to define what should be more important. By default, unnamed layers have the highest priority. Therefore, anything that is not yet in a layer will behave as it does today without using layers – the specificity will determine what is overwritten.

We are not restricted to mentioning or defining the layer only once. In fact, we can add things to a layer at any time. In addition, by default, every named layer is ranked in importance with its introduction time.

As a result, layers that are introduced later than others are prioritized before those. The escape hatch is to define the ordering ourselves. For instance, we could also write the previous code as follows:

```
@layer framework, site;

@layer site {
  .my-single_class {
    color: rebeccapurple;
  }
}

@layer framework {
  .overly#powerful .framework.widget {
    color: maroon;
  }
}
```

Now framework has a lower priority than site – applying the correct color for elements with the provided class.

One very useful option is to leverage the nesting capability and come up with an order like this:

```
@layer reset, microfrontend, app;
```

Now we can import related styles into the micro frontend layer:

```
@import url('../other/micro-frontend/style.css') layer(microfrontends.
other);
```

Importantly, the actual CSS code of the micro frontend remains untouched by this. Through the layer, we ensure that the micro frontend's priority is less than the styles coming from the application itself – with the exception of the general reset rules.

> **Important note**
>
> Modern CSS is an umbrella term used to describe recent advances in the platform that are either available on all browsers already, or on their agenda using something like the Baseline project. Each year the major browser vendors come together to align what features should be available on all platforms. For 2023, you can find the list of features at `https://web.dev/blog/baseline2023`.

Every method presented in this chapter is a viable contender for most micro frontend solutions. In general, these methods can also be mixed. One micro frontend could go for a shadow DOM approach, while another micro frontend could be happy with Emotion. A third library might opt for Vanilla Extract.

In the end, the only thing that matters is that the chosen solution is collision free and does not come with an unacceptable runtime cost. While some methods are more efficient than others, they all provide the desired styling isolation.

Consider an example micro frontend that was redone using some of the discussed techniques:

Method	Index [kB]	Page [kB]	Sheets [kB]	Overall [kB]	Size [%]
Default	1.719	1.203	0.245	3.167	100%
Convention	1.761	1.241	0.269	3.271	103%
CSS Modules	2.149	2.394	0	4.543	143%
Shadow DOM	10.044	1.264	0	11.308	357%
Emotion	1.670	1.632	25.785	29.087	918%
Styled Components	1.618	1.612	63.073	66.303	2093%
Vanilla Extract	1.800	1.257	0.314	3.371	106%
Tailwind	1.853	1.247	0.714	3.814	120%

Table 13.1: Evaluation of different CSS isolation techniques in an example micro frontend

In general, staying close to plain CSS will definitely keep the bundle size low and the performance high. Keep in mind that some of these techniques, such as styled components, can certainly be optimized more and would – if used for multiple micro frontends – not result in a much worse performance.

Furthermore, the performance impact depends largely on the implementation. For instance, for CSS-in-JS you might get a high impact if parsing and composition are fully done at runtime. If the styles are already pre-parsed but only composed at runtime, you might have a low impact. In the case of a solution such as Vanilla Extract, you would have essentially no impact at all.

For the shadow DOM, the main performance impact would be the projection or moving of elements inside the shadow DOM (which is essentially zero) combined with the re-evaluation of a style tag. This is, however, quite low and might even yield some performance benefits if the given styles are always to the point and dedicated to one particular component shown in the shadow DOM.

Summary

In this chapter, you learned how to properly isolate your styles to avoid conflicts on the presentation layer. While styling is one of the most important tasks when designing a web application, it also becomes one of the most challenging problems for micro frontends. Using the techniques outlined in this chapter you can now decide how to properly guard against style conflicts.

Ideally, your solution embraces a standardized way to use stylesheets with implied isolation. This way, you will not need to introduce additional tools or processes for verifying that styling remains intact with updates to the micro frontend configuration. Choosing a proper way forward involves thinking about the trade-off between flexibility and reliability, which can be quite tricky.

In the next chapter, we will look at how micro frontend solutions can be properly secured. This involves the identification of potential security vulnerabilities and attack vectors, as well as coming up with strategies to mitigate these.

14

Securing the Application

In the previous chapter, we saw how a set of techniques can be used to prevent micro frontends from leaking styles. While styling is certainly one aspect that needs to be considered to have separately created micro frontends all running in one application, the scripts and overall logic also need to be considered. Unlike CSS, there is no direct solution to the problem of missing script isolation.

In this chapter, we'll introduce the various security aspects that come into play when combining applications from multiple sources under a single runtime. Problems can be partially reduced by delegates and runtime isolation. However, in many cases, such techniques cannot be applied – at least for all relevant micro frontends.

One aspect that we want to investigate in this chapter is how micro frontends can be written more securely using web standards. In this area, we'll explore what can be done to limit the access from and to scripts originating from micro frontends. Finally, a key point in our journey will be to reliably verify the authenticity of content coming from the different micro frontends.

We'll structure this chapter as follows:

- Using web standards for hardening security
- Limiting script access
- Verifying authenticity

The samples shown in this chapter will be fundamental to ensure that any micro frontend solution created is at least as secure as its monolithic counterpart would have been.

Technical requirements

To follow the code samples in this book, you need knowledge of JavaScript and how to use the command line. You should have Node.js (version 20 or higher) installed using the instructions at https://nodejs.org. For this chapter, additional knowledge of HTTP and cryptographic algorithms could be helpful.

The code used in this chapter can be found in the following repo:

```
https://github.com/PacktPublishing/The-Art-of-Micro-Frontends-Second-
Edition/tree/main/Chapter14
```

For this chapter, there is no Code in Action (CiA) video available.

Using web standards to harden security

Security is one of the most important aspects when creating a web application. Whole companies have been victims of compromised security, so you can imagine what happens to single projects. To prevent such extreme scenarios, we need to keep a very close eye on potential security issues.

Quite often, security issues arise when the implementation of our solution deviates from the official standards. For instance, if for some reason we want to avoid session cookies as a means of authentication in our application, we could use a **JSON Web Token** (**JWT**) instead.

While the security aspects of a JWT deserve their own book, most security problems with this approach will come down to the question "*Where should it be stored in the user's browser?*". With a session cookie, this question would not be asked. The browser takes care of storing and providing it to the appropriate requests. The browser also restricts access to the session cookie as needed.

The authentication scenario is just one of many examples that come up when we talk about security. In general, we want to stick, as much as possible, to the web platform, which involves using the available web standards. This way, we get optimal security out of the box, as pretty much all web standards are designed to be as secure as possible.

Another example that is worth thinking about is the use of an `<iframe>`. As we've already discovered, in *Chapter 6*, these constructs come in handy for micro frontends in particular when we cannot fully trust the origin of the micro frontend. Surely, one could try to find all kinds of technically interesting ways to embed content somehow, but without the sandboxing features of an `<iframe>`, we will always fight an uphill battle.

One thing to definitely include in any web application is a proper **Content Security Policy** (**CSP**). A CSP allows us to define a whitelist of trusted sources for content loading, scripts, styles, and other resources. It helps mitigate the risks of **Cross-Site Scripting** (**XSS**) attacks by restricting the sources from which certain types of content can be loaded.

To enable CSP, we need to send the appropriate headers together with the app shell or the delivered HTML document. Alternatively, this can also be configured by having an equivalent `<meta>` tag in the document:

```
<!DOCTYPE html>
<html lang="en">
<head>
  <meta charset="UTF-8">
```

```
  <title>Example CSP</title>
  <!-- Content Security Policy -->
  <meta http-equiv="Content-Security-Policy" content="default-src
'self'; script-src 'self' https://cdn.example.com; style-src 'self'
*.googleapis.com; font-src 'self' https://fonts.gstatic.com;">
  <!-- Other meta tags, stylesheets, etc. -->
</head>
<body>
  <!-- Your page content here -->
</body>
</html>
```

By introducing the `Content-Security-Policy` meta information, we can set up the CSP for the document. Let's break down the example value that was set using the `content` attribute:

- `default-src`: This specifies where everything is allowed to originate from; the special value `'self'` denotes that all resources are restricted to come from the same origin, effectively preventing loading resources from external domains.

- `script-src`: This specifies where scripts can be loaded from, overriding the default source. In the example, we allow `'self'` (i.e., the same origin), as well as scripts originating from `https://cdn.example.com`.

- `style-src`: This specifies where style sheets can be loaded from, overriding the default source. Like the script source, we allow `'self'`, as well as a set of external domains using a wildcard (`*.googleapis.com`).

- `font-src`: This specifies where fonts can be loaded from, overriding the default source. Like the other definitions, we allow `'self'`, but add a specific domain (`https://fonts.gstatic.com`), too.

In general, the configuration for a CSP is highly dependent on the application. It does not only consider the desired security boundaries – which are usually only the same origin – but also what resources need to be loaded from where. In the preceding example, the stylesheet and fonts seemed to be delivered from Google Fonts, making those additional policies necessary.

> **Testing your policy**
>
> The CSP itself is quite strict – that is, if a request does not fulfill the provided policy, it will be denied. This can lead to all kinds of errors, which are quite annoying when introducing a CSP. A better alternative is to provide the `Content-Security-Policy-Report-Only` header with the value `policy`. This way, actual errors are treated as warnings – allowing us to develop our application, see what warnings appeared, and fine-tune the policy up to the point where we are quite sure that it works as it should.

Besides a proper CSP, two other things should definitely be considered to be introduced to have a secure micro frontend solution. One is **HTTP Strict Transport Security (HSTS)**. HSTS instructs browsers to always use HTTPS to connect to your server, reducing the risk of man-in-the-middle attacks and cookie hijacking. It is activated when a server responds to a request with an HTTP header called `Strict-Transport-Security`, indicating that the site should only be accessed using HTTPS for a specified period of time.

The primary goal of HSTS is to prevent requests to resources that could easily be abused for so-called man-in-the-middle attacks. With HSTS and CSP, we are quite safe already when the user interacts directly with our website. What might still be problematic is if our application is embedded in another website. While there are certain security headers, such as `X-Content-Type-Options`, `X-Frame-Options`, and `Referrer-Policy`, to mitigate various types of attacks and protect against information leakage, nothing beats the second standard that we should use in our micro frontend solution: **Cross-Origin Resource Sharing (CORS)**.

CORS allows us to define which origins are permitted to access resources from our micro frontends. By configuring CORS properly, we can efficiently prevent unauthorized access from other domains – essentially safeguarding our users from almost all potential phishing attacks.

For us to use CORS, we need to configure the server appropriately. Otherwise, the browser will just use its default CORS policies, which might prevent resource access under certain conditions.

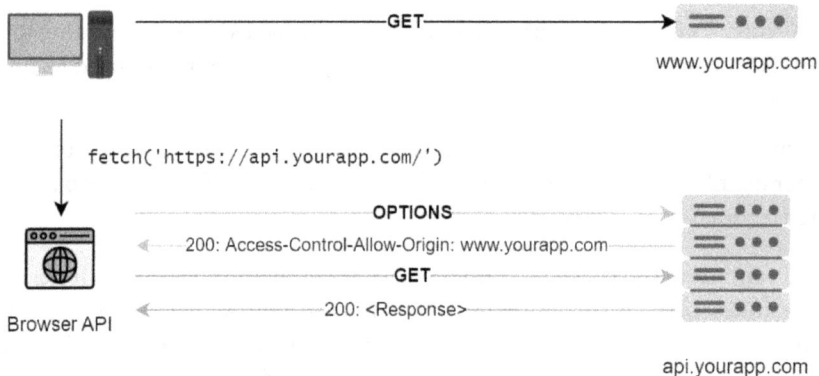

Figure 14.1 – Execution flow of a cross-origin API request

In *Figure 14.1*, we can see how CORS is supposed to work for a cross-origin API request. Any request from a website to a domain other than the origin will first have to perform a preflight request that is handled by the browser. Only if the information returned from the preflight request is evaluated to allow the actual request to be performed will the browser trigger the request. Otherwise, the original `fetch` call throws an exception.

As an example, when using Node.js with Express, we can establish a CORS middleware like this:

```
const express = require('express');
const app = express();

// Enable CORS for all routes
app.use((req, res, next) => {
  // Allow requests from any origin
  res.header('Access-Control-Allow-Origin', '*');
  // Allow specific headers
  res.header('Access-Control-Allow-Headers', 'Origin, X-Requested-
With, Content-Type, Accept');

  // Handle "OPTIONS" requests
  if (req.method === 'OPTIONS') {
    // Allow specific HTTP methods
    res.header('Access-Control-Allow-Methods', 'GET, POST, PUT,
DELETE, PATCH');
    return res.status(200).json({});
  }

  // Otherwise proceed with next middleware
  next();
});
```

In the preceding example, we add a middleware function to the Express app instance. In Express, this is done with the use function. A middleware function accepts three parameters: the request, req; the response, res; and a callback, next, to continue with the subsequent request handler. The implementation of the CORS middleware adds the appropriate headers for all requests but only continues with the subsequent request handler in the case of an ordinary request. Preflight requests using the OPTIONS method are directly handled.

With such a middleware, we are able to define what applications can request the resources from our server. Likewise, we can constrain what headers can be used – for example, we can use this to prevent sending the authorization header across different domains.

Once we follow the given web standards, it's time to reflect on our security model again. Since we include micro frontends from a variety of teams, we might have an issue with script files being evaluated and run without restrictions. Logically, we need a way to optionally limit the power of those scripts. That is what we'll learn to do in the next section.

Limiting script access

Any discussion about the security of micro frontends first requires a context to be defined. Are we talking about micro frontends composed on the server or on the client? Or both? As an example, if we talk about micro frontends following the island composition introduced in *Chapter 7*, then we have a scenario at hand where scripts from different micro frontends are evaluated on a central server.

The central server is certainly beneficial from a performance point of view, but also problematic in the security area. If we just imported the scripts from any team and treated them like all the other modules from the same server, we'd run into security issues such as arbitrary code injection, file system manipulation, or other unwanted risks.

The way out of this is to sandbox the scripts. In Node.js, we can do that by using the in-built vm module. This makes it necessary to specify the allowed global variables, including a `require` function to import other modules.

The following code achieves exactly that:

```
const vm = require('vm');

// Create a new context for the script
const sandbox = {
  console, // Allow access to console for logging
};

const context = new vm.createContext(sandbox);
// Evaluate the code "mfScriptCode" as a script
const script = new vm.Script(mfScriptCode);
// Run the script in the context
script.runInContext(context);
```

With the help of the vm module, we create a new script that is running in a new sandboxed context. Importantly, the sandbox environment is completely determined by us. In the example, we only allow access to the console using the `console` global variable.

Realistically, we'd provide a lot more than just the console – but certainly less than for our usual scripts. For instance, it's definitely not recommended to blindly pass in the `require` function. Instead, we'd want to pass in a proper wrapper around the `require` function that only allows certain in-built modules, or that returns mocked modules when importing another module – for example, the `fs` module.

The `mfScriptCode` is the text of the micro frontend's script to run – that is, the actual code as some other team deployed it. If the script contains `require` calls to other scripts within the micro frontend, we'd also need to fetch those. As calling the `require` function needs to be synchronous, this would not be possible; except if we preloaded the scripts.

A better alternative is to demand that only asynchronous imports are used. This way, we can still redirect the code to handle the asynchronous import in a way that aligns with our sandboxing approach – without having to preload anything.

While sandboxing on the server is possible, it becomes increasingly difficult for the client. Right now, there are only two options: either we place the micro frontend in an `<iframe>` and isolate everything or we run the script from a web worker – with a proxied browser environment. The latter, however, requires a lot of work and will come with a drastic performance penalty.

Finally, one last thing to consider for hardening the security of your micro frontend solution is to verify the authenticity of the micro frontends used.

Verifying authenticity

While a CSP and CORS are great for restricting the browser from fetching resources from unwanted domains, they are certainly not the solution for everything. One instance where the CSP itself is quite helpless is when the original sources have been modified to contain unwanted content. As an example, a script might unknowingly be altered to contain a keylogger that sends sensitive information to an untrusted domain. If the script is still served from our domain, we'd still load it – resulting in increased danger for our users. One way to mitigate this is to introduce integrity checks with the **Subresource Integrity** (**SRI**) standard.

In its primary function, SRI ensures that the resources, such as scripts or style sheets, loaded by our micro frontend components haven't been tampered with. It involves adding an `integrity` attribute to used script and link tags, containing cryptographic hashes of the resources they reference.

The following example shows the usage of the `integrity` attribute:

```
<!-- Example script with SRI -->
<script src="https://cdnjs.cloudflare.com/ajax/libs/jquery/3.6.0/
jquery.min.js" integrity="sha384-9a8e30fae59a71becea9d5e21a48b0c68b1a5
8a73b79f29c5f44b14c06f4d145"></script>


<!-- Example stylesheet with SRI -->
<link rel="stylesheet" href="https://cdnjs.cloudflare.com/ajax/libs/
twitter-bootstrap/5.3.0/css/bootstrap.min.css" integrity="sha384-
ZYSSSHVhfoekihjkgvjgjgghsgsfh1ks/fVIjckvdksfvdksf+vksdfg">
```

The first part creates a new `<script>` element from a third-party source. In order to ensure that the third-party source does not modify that script from the known version to a modified version with potentially malicious content, we tell the browser to check the script's content with the `integrity` attribute. The second part does the same – but instead of a `<script>` element, we use a `<link>` element referencing a style sheet.

The value of the `integrity` attribute uses a special format: It consists of two parts – a prefix denoting a hash algorithm and the actual hash value computed with the specified hash algorithm. For instance, the prefix `sha384` indicates that the SHA-384 hashing algorithm was used for the following hash value. The hash value is typically represented as a hexadecimal string.

> **Hashing algorithms**
>
> A hash algorithm can be used to map any content to a fixed-size value. The returned value is called a hash value. There are a number of well-established hash algorithms that are used widely. The **Secure Hash Algorithms (SHA)**, for instance, are widely used in security-relevant applications.
>
> When specifying the integrity of a resource using SRI, you provide the hash value generated by applying the chosen hashing algorithm to the content of the resource. In the example, the hash value represents the content of the script or stylesheet file retrieved from the CDN. The browser then verifies that the received resource matches the provided hash value, ensuring its integrity.
>
> Different hashing algorithms can be used with SRI, such as `sha256`, `sha384`, or `sha512`, depending on the desired level of security and compatibility requirements. In this case, `sha384` was chosen for its balance between security and compatibility.

SHA-384 is a member of the SHA-2 (Secure Hash Algorithm 2) family of cryptographic hash functions, which are designed to produce a fixed-size (384 bits in this case) hash value from input data of arbitrary size.

Verifying the authenticity of the used scripts is essential for providing a secure application. Boundaries that are set on the architectural and infrastructure level might be broken without runtime authenticity checks. Still, besides the security aspects that we have discussed in this chapter, we are always just as good or secure as the weakest link in the code base.

Therefore, security is a topic that starts at the local commit and continues to a central point, managing these micro frontends. A good gatekeeper to have is a discovery service, as we will explore in *Chapter 15*.

Summary

In this chapter, you learned how to strengthen the security of your micro frontend solution. You saw that embracing web standards and thinking about proper verification methods will help you mitigate most issues. While choosing the right architecture certainly yields the best possible foundation, certain limitations must also be installed in your solution.

The key point to take away from this chapter is that a solid security strategy starts with defining proper boundaries. Once you know the boundaries and potential attack vectors within these boundaries, you can outline the right takes to mitigate those risks. Always make sure to verify the origins of all assets used – especially for those that run into the user's or your infrastructure.

In the next chapter, we will look at how to make your solution even more scalable by using a micro frontend discovery service. While we have already touched on the topic in previous chapters, such as *Chapter 11*, by introducing a feed server, in the upcoming chapter, we'll look at such a service and its benefits in a lot more detail.

15

Decoupling Using a Discovery Service

In the previous chapter, we saw how micro frontend solutions can be made as secure as their monolithic alternatives. With this, we now know how to scale in terms of dependencies, styling, and security. What remains is an in-depth discussion of how to scale operations. This is where an architectural piece known as a micro frontend discovery service comes in.

In this chapter, we'll explore what functionality is provided by a micro frontend discovery service. We'll see what other capabilities might come with such a backend service. To best understand how such a discovery service works, we'll implement a simple one within this chapter. Finally, we will also use some advanced capabilities from a freely available third-party service.

We'll structure this chapter as follows:

- Avoiding hidden monoliths
- Implementing a discovery service
- Using advanced capabilities

With the ideas from this chapter, we'll be able to create solutions that allow instant onboarding, rollbacks, and A/B tests, among other things, making micro frontends not only a great organizational solution but also a fundamental technical enabler for efficient development.

Technical requirements

To follow the code samples in this book, you need knowledge of JavaScript and how to use the command line. You should have Node.js (version 20 or higher) installed using the instructions at `https://nodejs.org`. For this chapter, additional knowledge of the Node.js web application framework Express can be helpful.

The code used in this chapter can be found in the following repo:

`https://github.com/PacktPublishing/The-Art-of-Micro-Frontends-Second-Edition/tree/main/Chapter15`

For this chapter, there is no Code in Action (CiA) video available.

Avoiding hidden monoliths

Some micro frontend frameworks attempt to visually divide the UI, but that's not how real applications work. In reality, an application is a blend of various parts from different subdomains, converging to create a cohesive whole. While these subdomains can be neatly separated conceptually, they often coalesce within the same layout elements for the end user.

Although some micro frontend frameworks attempt to segment the UI visually, this alone is often insufficient. Real applications require that backend logic and data flow are also segmented in alignment with the UI to avoid hidden monoliths.

Consider a web store: product details and order history may reside in separate subdomains, yet users expect coherence. They wouldn't want to see just product IDs in their order history; they expect product names and relevant details. Consequently, these subdomains visually merge for the user.

Almost every subdomain contributes something to shared UI elements such as navigation, headers, or footers. Thus, creating a micro frontend solely for navigation overlooks practicality. Such a micro frontend would inevitably be overwhelmed with requests from other teams, turning it into a bottleneck. This setup, with its hidden monolithic tendencies, isn't ideal.

One might argue that excluding navigation from a micro frontend would merely shift the burden of changes to the app shell owner, as navigation is part of the app shell layout. However, this would exacerbate the problem.

So, what's the solution? Clearly, we need to decouple these elements. Instead of promoting strong coupling through direct imports with explicit relationships, we should opt for loose coupling. This means employing indirect imports with a fallback strategy.

Figure 15.1 – Strong coupling comes from the direct import of components across boundaries

In *Figure 15.1*, we see how strong coupling is formed by importing a component from another micro frontend inside the current micro frontend. Once strong coupling is established, it is impossible to treat the micro frontends as individual pieces. Consequently, the whole application becomes as tense as a monolith – not being able to adjust with the desired flexibility.

The goal should be to replace everything that can be interpreted as strong coupling between the micro frontends by loosely coupled mechanisms. Most importantly, we don't want to use the components from these micro frontends directly as it would introduce strong coupling. A good way to solve this puzzle is to introduce a wrapper component that is globally available. Through a globally available wrapper component, every micro frontend is capable of using all available components.

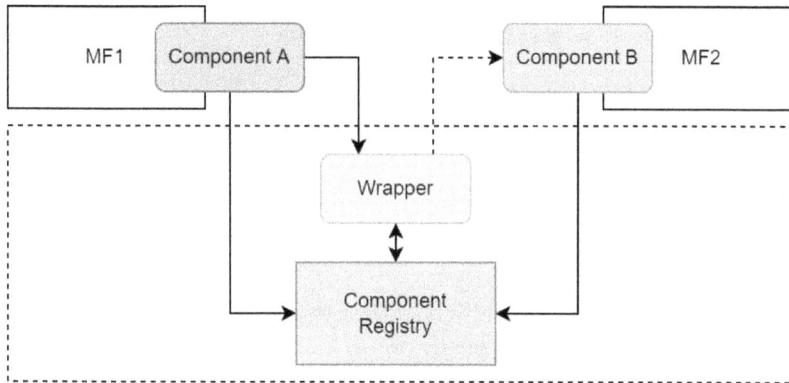

Figure 15.2 – Loose coupling can be achieved by using a globally available wrapper component

In *Figure 15.2*, we see how the globally available wrapper component can be used to replace direct imports of components from other micro frontends. Crucially, the wrapper component needs to work against an available component registry, which must be filled by the respective micro frontends. As an example, when **micro frontend 2 (MF2)** from *Figure 15.2* is loaded, it has to register **Component B** in **Component Registry**.

While the information that gets into the component registry is primarily filled from the micro frontends, it might be that the micro frontends only use an indirect approach to fill it. For instance, each micro frontend could provide a JSON that indicates what components should be registered under which name. Alternatively, as we've seen in *Chapter 11*, a special function that is exported from the micro frontend can be used to allow the micro frontend to register its components directly.

In any case, we need a source to obtain the collection of micro frontends to use. Importantly, this source then determines what components are registered in the locally available component registry. One way to create such a source is to look at the landscape of microservices to identify what they did to solve a similar problem.

In the microservices space, the solution to prevent tight coupling by gathering distributed information into a central source is called a discovery service. Let's explore what is necessary to build one ourselves.

Implementing a discovery service

A micro frontend discovery service is essentially a database containing information about the available micro frontends. The basic information for an individual micro frontend consists of the following:

- Name and version
- Some additional metadata such as author and description
- URLs to one or more files acting as entry points for the micro frontend
- The dependencies shared from the micro frontend – this should include their names, versions, and URLs to the files

This way, the registry not only aids in efficient component retrieval but also ensures dependencies are managed and updated correctly, crucial for maintaining system integrity and up-to-date interactions.

As an example, by calling the Piral Cloud Feed Service (`https://feed.piral.cloud`), which is a publicly available micro frontend discovery service, we get a response similar to the following snippet:

```
{
  "items": [
    {
      "name": "mf",
      "version": "1.0.0",
      "description": "Sample Micro Frontend",
      "dependencies": {
        "react@18.0.0": "https://assets.piral.cloud/sample/mf/1.0.0/
react.js"
      },
      "requireRef": "webpackChunkpr_123",
      "link": "https://assets.piral.cloud/sample/mf/1.0.0/index.js",
      "spec": "v2"
    }
  ]
}
```

In this kind of model, all the micro frontends are listed in an array in the `items` property. Choosing to enclose the array of items in an object makes sense for flexibility. Otherwise, if an array is returned instead of the object, we would not be capable of extending the response later on. With this approach, we can extend the response to add additional information without having a breaking change in the API.

Another option is to change the representation of the response. The Piral Cloud Feed Service comes with a wide selection of possible representations. One of the more popular options is the discovery service response following a proposed specification.

Discovery service response

The example showcased by the Piral Cloud Feed Service isn't the sole way responses can be represented. Various frameworks have devised their own notations, while others leave it to applications to decide. Additionally, a proposed standard for micro frontend discovery service responses is available. For deeper insights, delve into the GitHub repository housing the specification: `https://github.com/awslabs/frontend-discovery`.

Using the proposed micro frontend discovery service standard representation, we'd get the following response from the service:

```
{
    „schema": „https://mfewg.org/schema/v1-pre.json",
    "microFrontends": {
      "mf": [
        {
          "url": "https://assets.piral.cloud/sample/mf/1.0.0/index.js",
          "metadata": {
            "version": "1.0.0"
          },
          "extras": {
            "pilet": {
              "spec": "v2",
              "requireRef": "webpackChunkpr_123"
            },
            "dependencies": {
              "react@18.0.0": "https://assets.piral.cloud/sample/
mf/1.0.0/react.js"
            }
          }
        }
      ]
}}
```

The response comes with a schema property that makes it easy to reason about the provided structure by consuming the referenced JSON schema. The available micro frontends are specified in the microFrontends key, which is an object listing each micro frontend with its name as key and one or more versions in an array as value. This way, the format is already equipped to serve multiple versions of a micro frontend – where the client determines what version to actually include.

While url and metadata contain the core metadata, the extras property has been introduced to allow framework-specific metadata to be added by the discovery service.

From the point of view of a micro frontend consumer, this is enough to know about a discovery service. After all, a micro frontend consumer only cares about having an endpoint for getting a list of available micro frontends, as shown in *Figure 15.3*.

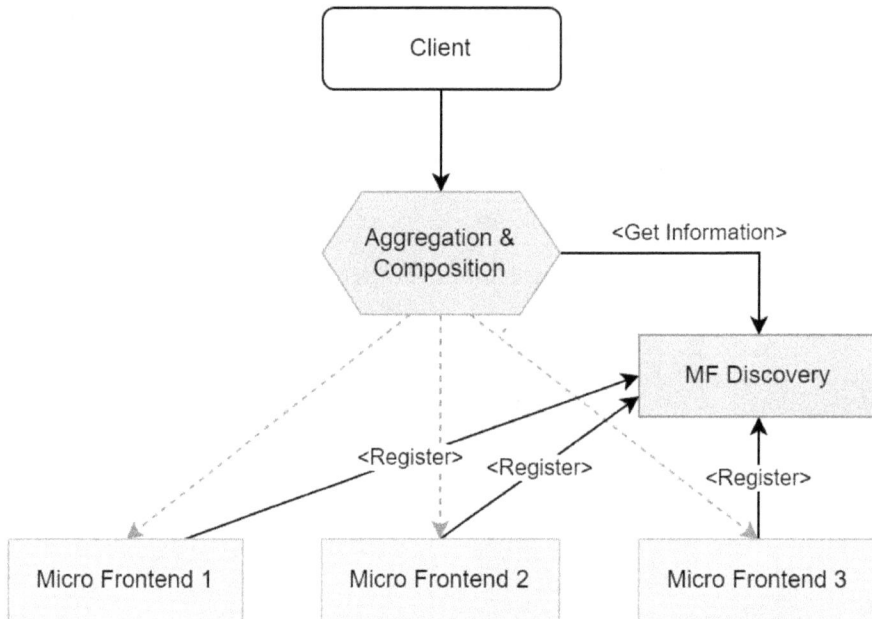

Figure 15.3 – A micro frontend discovery service serves as an information database

To be listed in the micro frontend discovery service, micro frontends need to register themselves first. This is shown in *Figure 15.3*. Any consumer of the available information can then get access to the URLs of the registered micro frontends. By using these URLs, it's possible to render an application – either from the server or the browser.

When we switch gears to the viewpoint of a micro frontend owner, there's a bit more to dive into. Here, the micro frontend discovery service is the gateway to publishing micro frontends. This allows teams to create tokens for publishing individual micro frontends. These tokens serve as security keys for producers to authenticate their publish request against the micro frontend discovery service.

To accommodate such a publish token, we can modify the code of the feed server that we developed in *Chapter 11*. In particular, we need to change the `publishModule` function to look as follows:

```
const allowedTokens = [
  ,f3c3564b-7e05-4db1-85ed-0cc54f7ea9dc',
];

exports.publishModule = (rootUrl) => (req, res) => {
  const bb = req.busboy;
```

```
if (bb) {
  const auth = req.headers.authorization;

  if (!allowedTokens.includes(auth)) {
    return res.status(401).json({
      message: 'Valid token required.',
    });
  }

  // as before in Chapter 11 - Siteless UIs
} else {
  res.status(400).json({
    message: 'Missing file upload.',
  });
}
};
```

In this sample code, we use a set of predefined tokens to limit the publish requests to those with a matching token. The token is supposed to be delivered in the `authorization` header. Keep in mind that for a real-world application, we'd connect to a database for storing and retrieving the available publish tokens.

In the case that no valid token has been found, we immediately return with an appropriate status code. Usually, a status code of `401` (unauthorized) or `403` (forbidden) is what we'd take here.

Now that we have established a solid foundation based on the feed server from *Chapter 11*, we'll look at some advanced capabilities that we could add to a micro frontend discovery service.

Using advanced capabilities

A micro frontend discovery service can flexibly point to different versions of a micro frontend, based on the evaluation of statically or dynamically provided rules. While this capability is certainly important, it is by far not the only feature that could be included in a micro frontend discovery service.

Take the Piral Cloud Feed Service, for example, offering the following:

- Dynamic rules driven by feature flags
- Custom configurations per micro frontend
- Detailed security reports
- Extensive analysis, including stylesheet insights
- Support for diverse formats, including non-JavaScript options

- Fragment management within micro frontends

- Aggregation of micro frontend collections

- Error reporting for individual micro frontends

- Usage statistics tracking

What benefits arise when such features are considered? For instance, implementing dynamic rules driven by feature flags allows administrators to control which versions of micro frontends are served to different user groups or under different conditions. This capability is essential for conducting A/B testing or for gradually rolling out features to users, thereby minimizing disruption and enhancing user experience.

Another example is that detailed security reports provided by the discovery service offer insights into vulnerability assessments, dependency security, and compliance status across all micro frontends. These reports are vital for preemptive security measures and for ensuring that the entire frontend ecosystem adheres to the latest security standards.

By offering aggregated collections of micro frontends, a discovery service can enable developers to manage dependencies and interactions through a centralized dashboard. This aggregation simplifies updates and integration, ensuring that components can effectively communicate and function as a cohesive unit.

A discovery service can conduct extensive performance analyses, including stylesheet usage and optimization insights. This helps in identifying inefficient CSS that may affect page load times and provides recommendations for optimizations that enhance both speed and visual consistency across micro frontends.

When a micro frontend discovery service supports a variety of frontend formats, it may extend beyond traditional JavaScript-based applications to include WebAssembly, server-rendered HTML, and even containerized micro frontends. This flexibility ensures that the architecture can adapt to a wide range of technologies and usage scenarios.

Furthermore, individualized error reporting for each micro frontend allows developers to quickly identify and address issues specific to each component. This granular level of insight is crucial for maintaining high uptime and ensuring that errors can be isolated and rectified without impacting the entire system.

Depending on which of these capabilities we desire, we'd need to either spend additional effort on their implementation or consider integrating third-party services. However, with every new capability, we definitely make the discovery service more powerful – and more useful, too.

Consider adding the feature of dynamic rules for provisioning user-tailored sets of micro frontends. This way, one user gets some micro frontends, while another user might get completely different micro frontends.

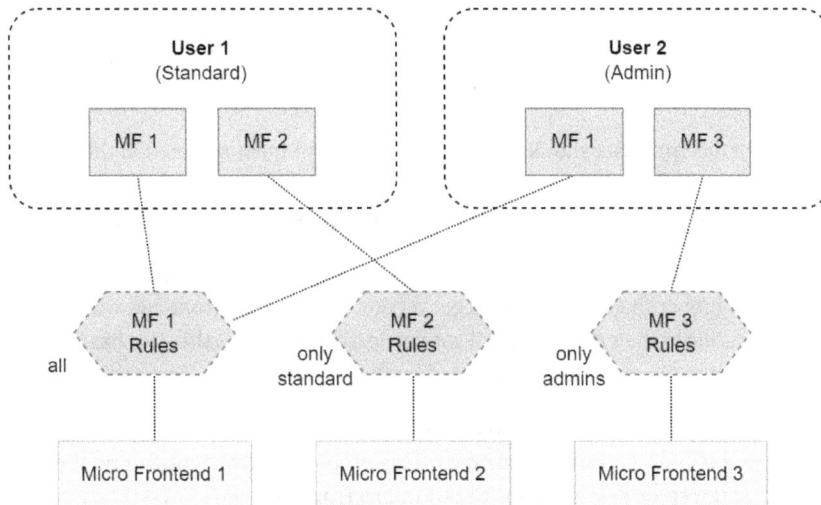

Figure 15.4 – Providing user-based sets of micro frontends with a rule engine

In *Figure 15.4*, we see how a discovery service could create user-based sets of micro frontends. Depending on the evaluation of the provided rules, some micro frontends are provisioned, while others will not be seen. As an example, micro frontends that are only relevant for administrators should not be seen by standard users.

In general, the rules are quite flexible. Anything that can be read by a web service could be used to determine what rules are possible. On a lower level, this implies reading URLs including their query parameters, as well as all provided standard and custom HTTP headers.

On a higher level, we can get extensive user information via IP tracking, authorization token inspection, or user-agent analysis. The provided information can also be used for some fingerprinting (e.g., to roll out some A/B tests). Obviously, the most reliable source for A/B testing is not fingerprinting, but an actual user session or user identification cookie.

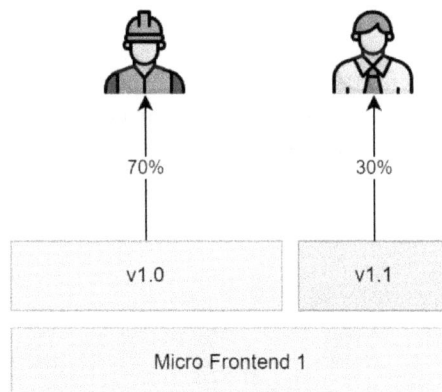

Figure 15.5 – A/B tests deliver a specific version of a micro frontend to a target user group

In *Figure 15.5*, we see how A/B tests can be deployed using a micro frontend discovery service with an integrated rule engine. Thanks to the clear identification of a user through the provided headers, we can assign an integer number from 0 to 100 for each request.

Subsequent requests of the same user should get the same integer number assigned. If they got a different number, the consistency would be broken – the same user would get one set of micro frontends in the initial request and another set in a later request. From the perspective of the user, the whole UI would become unpredictable.

Requests with a lower number than the threshold get version A of a micro frontend, while requests with the same or a higher number than the threshold get version B. This splits the user base into two parts – with the fraction of users for each side determined by the threshold set to serve the micro frontend.

A/B tests are not the only feature we get essentially for free when we add a rule engine to our discovery service. Another thing is blue-green deployments. This type of deployment can be thought of as a gradually changing A/B test for new versions of our micro frontends.

> **Blue-green deployments**
>
> Blue-green deployments originated from backend development to avoid downtimes. In that setting, the original (or current) instance of a service was kept alive while a new version was deployed and ramped up. Then, traffic was slowly redirected to reach the new version until the current version had zero requests and could be shut down.

In case of unclear runtime characteristics, a new version of a micro frontend would not be deployed instantly to all users, but rather to a small subset that is gradually expanded to the full user base. This way, we can stop more smoothly in case of severe errors – without affecting the whole user base.

There is, of course, much more we could dive into. The crucial point is to realize that certain functionality could be covered by a discovery service. For now, let's switch gears and summarize what we learned in this chapter.

Summary

This chapter detailed how a micro frontend discovery service not only supports the scalability of applications but also simplifies the management of micro frontends. By enabling dynamic onboarding, rollbacks, and A/B testing, the discovery service plays a pivotal role in maintaining an agile and responsive frontend architecture.

By introducing a micro frontend discovery service, you ensure that your solution remains scalable and flexible. Having a dynamic composition of your frontend, as outlined in *Chapter 11*, micro frontend discovery enables that. Rolling back and dealing with A/B tests in production? This is where a micro frontend discovery service really shines. Tracking back which team published what version of their micro frontend at what point in time? Luckily, a micro frontend discovery service holds the right database for these queries.

In the upcoming chapters, we'll see how we can efficiently deal with updates to the app shell from an organizational point of view. We will see that governance and design awareness are required to establish a common foundation.

In the next chapter, we will continue the journey of onboarding everyone – not focusing exclusively on developers – with a preparation round involving the major stakeholders. We'll discuss how everyone can be made aware of the technical principles, leading to improved expectations and more clearly defined requirements.

Part 4:
Busy Bees – Scaling Organizations

In this part, you will learn how to successfully manage micro frontends from an organization's point of view, including common pitfalls, communication strategies, and the impact on UX/UI. Finally, a set of case studies will provide you with insights from other projects.

This part covers the following chapters:

- *Chapter 16, Preparing Teams and Stakeholders*
- *Chapter 17, Dependency Management, Governance, and Security*
- *Chapter 18, Impact of Micro Frontends on UX and Screen Design*
- *Chapter 19, Building a Great Developer Experience*
- *Chapter 20, Case Studies*

16

Preparing Teams and Stakeholders

In the previous chapters, you've learned a lot about the existing patterns to implement micro frontends, as well as ways to improve their technical implementation and get around certain limitations. In all cases presented, the problem statement needs to be really well understood to determine the right architecture and introduce a sound and sustainable implementation.

Even with all the technical decisions made correctly, the project may still fail. In this chapter, we'll try to find out why projects may fail due to non-technical issues. Specifically, we'll identify what can be done to prepare teams and stakeholders to avoid misalignment and communication issues.

In this chapter, we cover the following topics:

- Communicating with C-level stakeholders
- Handling product owner and steering committees
- Team organization

> **Important note**
>
> Not all sections of this chapter may apply to you. Depending on the company and project, the responsibility of team organization or product owner communication might be given to somebody else. This is fine as long as everyone involved in the project is on the same page and knows how to fulfill their role in a setup leveraging micro frontends.

Now let's take the chance to dive into the topics.

Technical requirements

To follow the code samples in this book, you need knowledge of JavaScript and how to use the command line. You should have Node.js (version 20 or higher) installed using the instructions at `https://nodejs.org`.

For this chapter, there is no source code and no **Code in Action (CiA)** video available.

Communicating with C-level stakeholders

C-level stakeholders do not care about micro frontends. They also don't care about microservices – or other parts of the technical realization. Instead, they care about their business and the impact of technical implementations on their business.

It is easy to think that in a setup where the executive level does not care about technical implementations, you should not mention micro frontends. While the name *micro frontends* may not ring a bell for these stakeholders, you should still remind them about the actual benefits and pitfalls of the system at hand.

In the following two subsections, we will go into two important tasks when dealing with the executive level: managing expectations and writing executive summaries. We'll start with a brief excurse on expectation management.

Managing expectations

One of the most important things when writing executive summaries or performing C-level communication is the art of expectation management. Expectation management describes the way that progress and outlook of a project are articulated such that involved parties do not fall into the trap of either missing the opportunities at hand or actually believing that there will be more possibilities than given. In both cases, the result is a failed project or goal. Either way, there is a joint disappointment.

Consider including tools or methodologies for expectation management, such as **SMART** goals, regular status updates, and risk management frameworks. The SMART is short for *Specific, Measurable, Achievable, Relevant*, and *Time-Bound*. However, expectation management is not only restricted to communication with stakeholders, but also with other parties and the project team itself.

There are also multiple things that require proper expectation management. Among them we find the following:

- Company vision
- Business targets
- Project goals
- Feature intentions

Surely, most of them have nothing to do with a specific project, but should usually at least be aligned somewhat with the project.

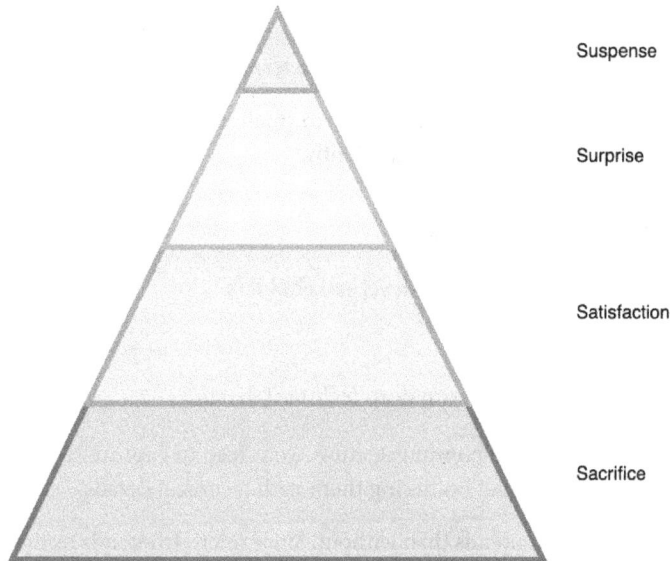

Figure 16.1 – The pyramid of expectation

As shown in *Figure 16.1*, there are multiple levels when dealing with expectations. While many just aim to satisfy an expectation, it may be better to surprise the audience. You might even go beyond this and build up a kind of suspense – where the audience is even anticipating an experience.

> **Important note**
>
> One of my favorite descriptions of expectation management comes from Blake Basset, a product manager at Microsoft Azure. He describes expectation management as a *multi-party affair, one that requires building a shared understanding with your team and across stakeholders about where you are headed, how you plan to get there, and what you will deliver. And that is just the beginning. Once established, expectations must be continuously reinforced and adjusted when needed.* Quite importantly, he emphasizes continuous effort.

To avoid a disconnect between the executing and the developing parties, you need to highlight the boundaries of the system. For instance, micro frontends should allow you to implement and release new features faster. Nevertheless, that does not imply that the software is bug-free or implemented instantly. It just means that the implementation can be done with fewer constraints and that the feature can be released pretty much after the work has been accepted.

Mentioning terms such as micro frontends, distributed development, frontend modules, or portable components may confuse executive stakeholders. Instead, we should use analogies highlighting the ins and outs of the system. As an example, instead of calling out distributed development, you can call it *independent work*. Bringing an analogy such as *interior works in a house, where each room can be done independently and in parallel* helps to specify the terminology without getting too technical.

Since most communication with C-level will be done in small written summaries, it makes sense to also review how you can compose them most efficiently.

Writing executive summaries

There are two things that seem inherent in C-level stakeholders:

- They are notoriously short on time
- They don't want to be concerned with deep technical details

Therefore, you need to be efficient in communication. You need to capture the gist of the current situation as quickly as possible without bothering them with technical details.

Luckily, it is simpler with micro frontends than without. Since micro frontends technically allow us to spend the development time on features rather than integration, you should not need to report blockers or technical impediments. However, once technical issues come up, it is important to focus on the solution rather than the issue; that is, don't fight why the problem occurred, but rather think forward to find a good way around it. This way, stakeholders will be happier and teams will be more productive.

A good executive summary allows the reader to digest quickly what is needed to go forward. Are more people needed? External consulting? Can software be purchased to help here? If the problem needs to be understood to make a decision, then the full report is here to help. If a full report is missing, somebody will come to ask for it.

As the technical lead for the whole micro frontend solution, it may be difficult to capture what is really going on. After all, there will be multiple streams with independent development going on. Here, you should not forget that even though development is distributed, the overall business – at least the one to report to – is still monolithic.

There are two solutions for gathering the status of distributed development:

- Request the status for each team; that is, get an executive summary/report from each team
- Place the responsibility of writing the report on a horizontal team or person involved in all development teams

The latter makes the most sense if there is a scrum master or agile coach guild that is shared among the teams. This guild can then gather to discuss progress and work out the report. Otherwise, the classic approach of distribution and aggregation can be used.

The executive level may be the most important level with respect to the business processes, however, even more important are the people directly in charge of the project. These project leads and product owners may even introduce additional circles of responsibilities quite often labeled as *steering committees*. Handling these parties is therefore at least as important for overall expectation management. Let's see what we should do about them.

Handling product owners and steering committees

Depending on your role, you will need to handle product owners and related entities such as steering committees. Maybe you are a product owner yourself. In any case, as with C-level stakeholders, there are certain communication rules that should be respected.

For product owners, one of the most important aspects of a project is to know who is responsible for what. In the end, tasks can only be efficiently fulfilled if they are dispatched to the right person. In a monolithic setup, the roles and responsibilities may be much easier to see than in a distributed organization. Obviously, we need to have a way of bringing in transparency.

One way of being more transparent in a micro frontend solution is to leverage techniques such as a **Responsible – Accountable – Consulted – Informed (RACI)** matrix. Each development team assigns one person who is responsible for filling in and maintaining the RACI matrix.

Responsible   Accountable   Consulted   Informed

Figure 16.2 – The RACI model to assign each role one of four levels per task

As shown in *Figure 16.2*, there are four levels that can be assigned per task. A task in the micro frontend space can be also related to a feature area. As such, tasks such as *Maintenance of the balance micro frontend* are possible. Not every person needs to be assigned one of these four levels – as not everyone is responsible, nor accountable, nor consulted, nor informed. Quite often, no level should be assigned to a role. A technical role, for instance, quite often only implies engineering tasks.

Another approach that has become more popular recently is **swarming**. Since swarming works by utilizing a collaborative network-based approach, it is an ideal fit for distributed development. Also, in comparison to the more conservative tiered support, which is measured by activity, swarming measures by value creation.

Conservative models usually tend to create silos and hierarchies. While there are certainly some benefits, such as a clear direction and an easy-to-understand linear pipeline, it fails to properly capture the dynamics of modern projects.

Swarming only works if the buy-in from the executive level is fully given. Since swarming offers some immediate drawbacks, such as mixing different pay grades heavily, the first reaction may be repulsion. It is quite likely that C-level will only see costs and not benefits. Also, the workflow will seem more chaotic. After all, in swarming, people come together dynamically to use pair programming and other techniques to solve the issues at hand.

To avoid getting lost in chaos, more scrum masters or agile coaches are needed. Consequently, the immediate cost will go up once more. However, in the end, the interaction between the product owner and the development team should be more frequent, too. While this may seem like an issue at first, it will lead to better software. If the requirements are not only fully understood, but also potentially improved, shaped, or questioned by the developers, then the quality of the resulting software is better.

Unfortunately, especially for larger projects originating in enterprises, another entity to take care of may appear: a steering committee. This is a circle of people to provide guidance, direction, and control to a project. While some steering committees are only formed to alleviate the responsibilities of the project lead, some committees are full control organs, which need to be consulted before doing any work – and especially before spending any money.

Quite naturally, a steering committee is the natural opponent of non-hierarchical work done, for example, in a swarming approach. The goal, however, should not be to fight the steering committee at all levels. Even though that may seem like the right thing to do, it will actually have the opposite effect. Instead, this means that the project will require more governance levels.

We will touch on the whole topic of governance in *Chapter 17* in full detail. For now, all we need to know is that we will need one central team to handle the steering committee and take care of all project aggregation duties. Usually, this team is also responsible for the app shell.

While dangerous for the success of a micro frontend – or any kind of modular solution – system, steering committees can also provide value if handled correctly. As we've seen with the different patterns, there are architecture styles that enable more freedom and ones where more decisions are made by the app shell already.

For instance, the Siteless UI pattern allows us to define multiple boundaries to constrain what the individual modules can do. Depending on the project and organization, this may be used to either give more or less freedom. Theoretically, it could be used to pre-determine most decisions excluding the covered domain-specific functionality.

In the end, this is all a question of the right team organization. Let's investigate what team setups are possible and how we can change existing organizations.

Team organization

Very often, adopting a micro-frontends style is not so much about the technical challenges, but rather about the organizational shift. Like microservices, the organization needs to embrace the architecture and reflect it in its team structure. The reflection of a business's organizational structure within its products and projects is also known as Conway's law.

> **Important note**
>
> Computer programmer Melvin E. Conway made an important realization in the late 1960s. He observed that *any organization that designs a system (defined broadly) will produce a design whose structure is a copy of the organization's communication structure.* His idea originated from the observation that different parts of a system will be done by different teams or different people, however, all these need to communicate with each other – thus just mirroring the organizational setup that they already found at the beginning instead of introducing new paths.

Consequently, a monolithic company culture will never fully embrace the distributed work of a micro frontend solution. Obviously, the easiest path would be to radically change the existing team setup, however, from experience, this rarely works. If it did work, the team's setup would have been changed before micro frontends were introduced – not afterward.

First, let's try to understand the team setups that are possible, thanks to micro frontends. Then we'll see how existing team organizations may be changed.

Understanding possible team setups

One of the best things about micro frontends is that utilizing this architecture allows almost any team setup. But as you know, with great power comes great responsibility. Choosing the right team setup is an art of its own.

What are the most popular setups? Obviously, by far the most popular setup is to just place one dedicated team per micro frontend. This is also known as the vertical team setup.

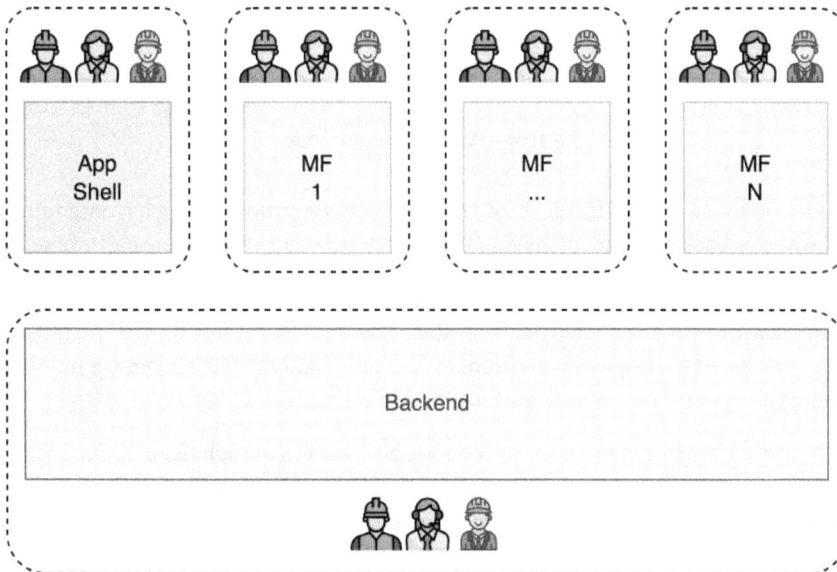

Figure 16.3 – Vertical team setup

Figure 16.3 shows the vertical team setup. Each team is fully independent and contains each role. Therefore, no team will be blocked, for example, with missing UX designs. In the illustration the backend is shown as one b

lock, however, that should only be regarded as a placeholder. The backend may be designed quite like the frontend, too, hosting multiple teams.

We can also envision an overlap between the backend and the frontend. This way we could create real full stack teams. Full stack teams are teams that have responsibilities in both frontend and backend code.

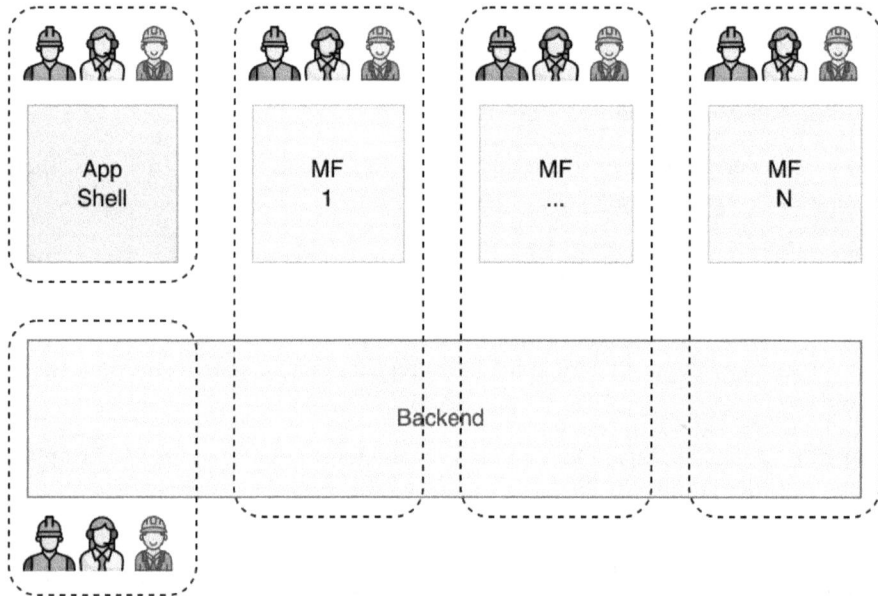

Figure 16.4 – Full stack team setup

As illustrated in *Figure 16.4*, having full stack teams does not exclude having a frontend team that does not deal with the backend – or still having teams that are exclusive to some part of the backend.

Naturally, the app shell is a frontend component that does not necessarily spawn a full stack team. Quite often, full stack teams are domain-driven, such that a full stack team owns the frontend and backend module assigned to their problem domain. Besides frontend and backend, a full stack team should also have full operational control, making it truly autonomous.

Going one step further, we might not want to replicate all the roles in each team. Some cross-cutting concerns such as quality assurance or UX design could be done by shared resources. These people would then work across the different teams. This is where the shared team setup comes in.

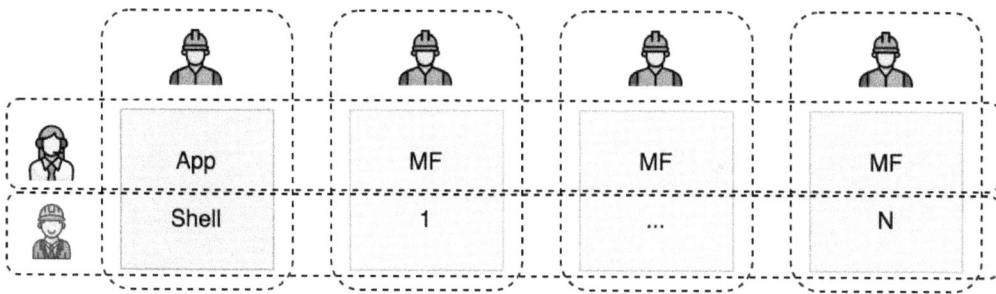

Figure 16.5 – Shared team setup

A diagram of a shared team setup is shown in *Figure 16.5*. Note that the shared resources are a cheaper approach at first but will lead to actively blocking some development time and thus wasting some money, too.

Overall, the shared setup makes the most sense when the demand for consistency is quite high. Quite often, the number of available people for a dedicated role is not the deciding factor to favor this approach.

The other extreme is certainly to share all development resources. This creates a horizontal setup, where every micro frontend is essentially developed by everyone. In practice, this approach may seem less common but is actually fairly popular, too. Unfortunately, it usually leads to problems.

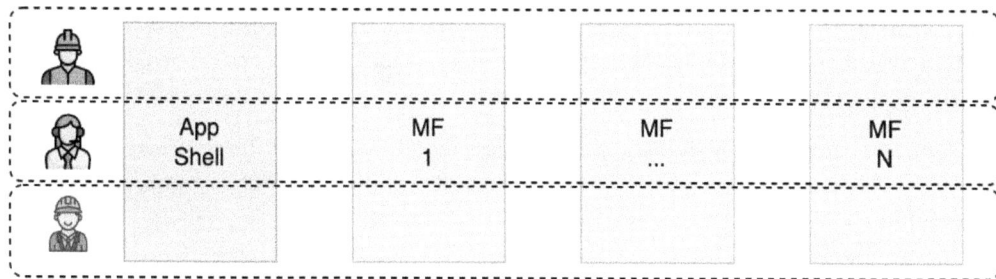

Figure 16.6 – Horizontal team setup

While there may be good reasons to choose this setup, it is very often an indicator that the move towards micro frontends was premature and that either the domain decomposition is insufficient or that the developer mindset is still entrenched in the monolithic approach. In *Figure 16.6*, we see the horizontal split shown schematically.

Note that in reality, we will not see every developer working on every module, but you get the gist of it.

Finally, we should not forget that the ideal solution may be a combination of the presented ones. We could refer to it as a *hybrid team setup*, which denotes that multiple philosophies are used to define the team organization.

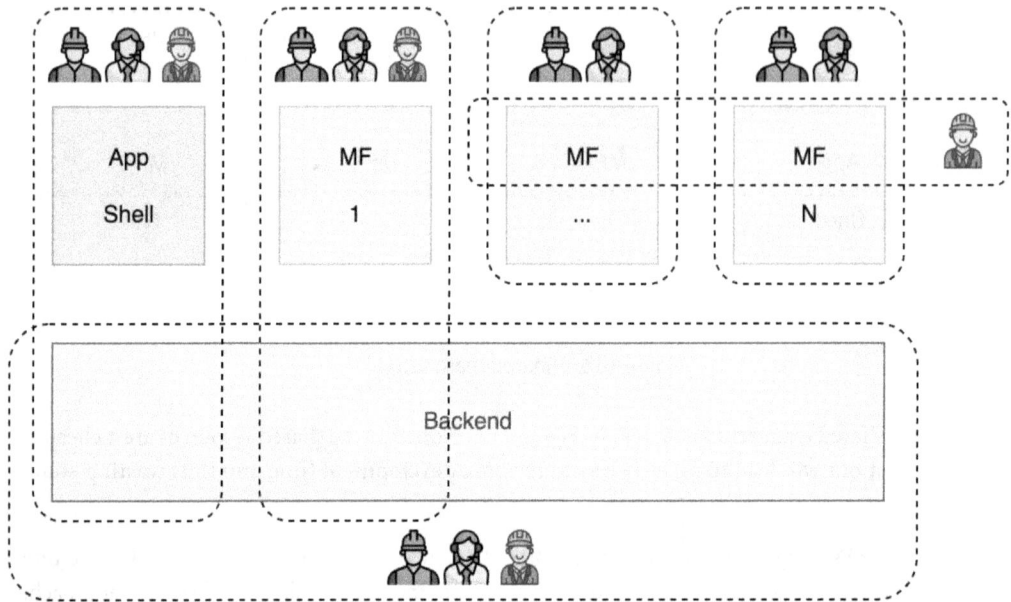

Figure 16.7 – Hybrid team setup

In *Figure 16.7*, we see one example of how a hybrid team setup may look. While some teams receive full stack responsibility, other parts of the application are maybe too small to justify a full team working on them. Here, either part is handled by horizontal teams or actually helped out by shared resources. Other parts may leverage autonomous vertical teams.

Quite often, the issue with the hybrid approach is less that it wouldn't fit the actual solution, but rather that it's difficult to efficiently consider so many custom setups. In the standard case, the domain decomposition was already done in a way to support a certain team setup.

In the end, the actual team setup does not matter so much. The important part is that the team members feel empowered and take responsibility for their work. Once a team setup enables this way of thinking, we are on the right track.

It's not very often we'll find the right team setup before we start. But even once we've settled on a certain team setup, we'll find room for more optimization. Let's see how we can change the team setup to reflect the desired organization principles.

Changing team organizations

Changing the setup of an established team is hard. Quite often, a much easier way is to end the team and assemble a new team using the new principles. This does not mean that people need to leave the project or company – quite the opposite. It just means assigning everyone to a role that fits within the new construct.

One strategy to successfully transition from an existing, potentially not-so-well-working team organization to a good setup is to merge and divide. First, we create a larger team consisting of existing teams. After a while, you need to be ready to divide into two or more smaller teams. The teams should be assembled appropriately, putting developers that like more monolithic structures into larger teams that are closer to the app shell or other features that naturally are larger and more structured in development. This way you can recombine developers until a good setup has been found.

One thing to remember when recombining is that a change in team organization will block development resources. Therefore, the efficiency will definitely decrease for some time. This time, which might be referred to as the transition period, is not a good representation for justifying whether a team setup is ideal or should be changed again. The iterative process of finding a good team setup is one of the slower processes.

Another way to change an existing team setup is to introduce additional faces with the capability of dragging everyone in the right direction. This could be either an agile coach, an architect who fully fits into the technical mindset, or a developer who believes in the planned solution.

Finally, to ensure that everyone is on the same page, use the following checklist:

- Regular alignment meetings to discuss platform strategy and changes
- Documentation is distributed and aggregated centrally
- Online communication (chat, issue tracker) between the different teams is established

With this, we've come to the end of the chapter.

Summary

In this chapter, you learned that micro frontends not only present a technical challenge, but also an organizational one. You've seen that there are various aspects that should be considered when communicating with the executive level, steering committees, or product owners. You now have a deep understanding of potential team setups and how to change teams efficiently.

Every team is unique. So, while some of the techniques shown here may apply perfectly to your team, others will not, or only in a modified form. Try to align with your team first. Find out which approach best fits your case and get on the same level of understanding. The path towards micro frontends can only be walked together as a team – never force it.

Ultimately, every organization adopting micro frontends will need to incorporate some changes or the project may deviate from the original plan in the best case, or in the worst case, be a complete failure.

In the next chapter, you'll see where the organizational and technical aspects overlap and what you need to take into account in order to be successful.

17

Dependency Management, Governance, and Security

In the previous chapter, you learned that micro frontends are as much an organizational challenge as they are a technological challenge. In *Chapter 2*, you read that dependency management, the governance of modules, and the security of an application are among the most difficult challenges that you can face when dealing with micro frontends. In this chapter, we will discuss these challenges in more depth.

Quite often, there are multiple ways of solving or at least mitigating a problem. It mostly depends on two factors: what architecture has been used for the micro frontend solution and how open or closed the system should be. Many of the concerns discussed in this chapter can be drastically alleviated by having direct access to and control over each micro frontend's source code.

In addition to the security aspects, many of the problems—or their direct solutions—also offer challenges in terms of performance or **User Experience** (**UX**). There is no silver bullet; however, being aware of the challenges, their potential solutions, and their properties is crucial for delivering an appealing micro frontend application.

In this chapter, we will cover the following topics:

- Sharing all or nothing
- What about change management?
- Establishing a governance model
- Sandboxing micro frontends
- General security concerns and mitigations

As discussed in *Chapter 12*, sharing dependencies is one of the crucial things related to micro frontends. This topic alone requires many discussions; this is because the two extremes, namely, sharing no dependencies and sharing all dependencies, don't age well either. Finding the perfect middle ground and being able to support that mode is crucial. Let's examine why this is the case and look at how we can find that sweet spot.

Technical requirements

To follow the code samples in this book, you need knowledge of JavaScript and how to use the command line. You should have Node.js (version 20 or higher) installed using the instructions at `https://nodejs.org`.

For this chapter, there is no source code and so no **Code in Action (CiA)** video is available.

Sharing all or nothing

We've touched on this point a couple of times already: should we share all dependencies or no dependencies? Coming from a microservice background, you might tend to share nothing, and you will have several good reasons for doing this. After all, sharing anything will lead to constraints and potential bugs.

On the other hand, if you need to include all dependencies in every micro frontend, the whole solution could become bloated and quite slow. It will also be unable to leverage some of the more advanced communication patterns.

This is another example of where the truth lies in the middle. Now the question is, how should you decide what dependencies to share? Generally, you should not share any dependency. Going with this default choice makes your life much simpler. Also, making a mistake by not sharing a dependency hurts less than making a mistake by sharing a dependency.

Great! Now that we have settled on a default choice, we still need to figure out a way to determine what dependencies to share. Let's start with some basic criteria:

- They should be of a significant size (from at least 15 kB to 30 kB; even more appropriate is 100 kB) following how fast the data transfer is for your average primary target user
- They should be used by at least two micro frontends
- They should not have multiple commonly used versions with breaking changes
- They should play an important role in rendering components
- They should be technical, not domain-specific

The last point, in particular, could be debatable. However, if we think in terms of what we have already learned about proper domain decomposition, we could guess that the need for sharing domain-specific modules will be an indicator of an insufficient domain decomposition. With such fundamental flaws in the application's architecture, we should first try to improve this decomposition before adding more shared dependencies.

One thing to keep in the back of our mind is that once a shared dependency is added, it might be very hard, even impossible, to remove. Consequently, it's one of the things that just leads the system to grow in size. Bloated applications are one of the reasons why massive rewrites are popular in the first place.

So, how can dependencies be shared? In *Chapter 12,* we've already seen that tools such as Webpack Module Federation can help us tackle this problem. Bundlers, in general, allow us to exclude dependencies. In webpack, this can be done by declaring desired packages as `external`, telling the bundler to exclude a certain dependency from the final output. At this point, there has to be some common knowledge between the sharing party, which is usually the app shell, and the consuming micro frontend. An easy way to do this is to just add the shared dependencies to the global `window` object.

However, an even better way of sharing dependencies can be introduced by allowing micro frontends to exchange dependencies freely—if they match. For details, see the concepts and associated examples introduced in *Chapter 12.*

Of course, the downside of sharing dependencies from micro frontends is the increase in complexity. Furthermore, the actual storage size of the micro frontends also increases. This is because they never know which other micro frontends are active and what shared dependencies are already available for consumption. However, the end result is presumably the best of both worlds: we get more flexibility and better performance.

Let's say we introduced a shared library centrally, but we want to update it. What can we do about it? How should we approach the problem? This is where active change management comes in handy.

What about change management?

With micro frontends, the team owning the app shell becomes a service provider. This is because any changes to the app shell will need to be announced and rolled out with sufficient communication upfront. Welcome to the world of change management!

Usually, change management refers to the impact on people and teams following an organizational transition. However, in software—or IT generally—change management goes through phases from planning to review to implementation.

Introducing a change management process is motivated by two well-known facts:

- Used applications need to change; otherwise, they become obsolete

- Complexity is never reduced; with every change, a system will become more complicated

Every change needs to start with a clear requirement. This does not necessarily need to be a new feature—it can also be a direct change request or a bug report. In any case, we need to identify the potential change and determine its technical feasibility.

Once we know what needs to be changed and how it will be implemented, the costs have to be estimated. This is necessary to align with the expected benefits and problems. If the expected benefits are low, the change should only be considered if the costs and risks are low, too. If there is a bigger advantage, the change might be implemented despite the higher costs and risks.

When we have decided whether a change should be implemented or not, we can go ahead and make a roadmap, as follows:

- Decide when and how to initially roll out the change (for example, should it be rolled out in waves starting with dedicated beta users?)

- Decide when to make the first announcement

- Decide when to make the final announcement

- Decide when (and how) previous functionality (if any) is to be deprecated or removed

Aside from direct user communication, there are many indirect ways to communicate:

- Utilizing a changelog

- Updating the documentation

- Making a new release

- Adding to the migration guide

Keep in mind that users will most likely be part of the development teams for the different micro frontends. There can—and will—be scenarios where a user is indeed an end user of the application itself. This is the case when the changed functionality has a direct end user impact. Of course, many of the described points still apply; however, in this case, the change management will most likely be marketing-driven, not technology-driven.

In the end, it all boils down to setting the right boundaries early on. Here, a proper governance model helps. Let's find out how to establish this next.

Establishing a governance model

The whole purpose of governance is defining processes to simplify everything that is happening and that is necessary for the product. This starts at the initial required analysis stage and ends with maintenance efforts.

From a technical point of view, a governance model helps us identify what processes can be left unautomated and which ones need automation.

In general, there are multiple areas that can be covered with such a governance model, as follows:

- Business analysis (such as roadmaps, timelines, and monetarization)
- Requirement definitions (such as use cases, effort, and risks)
- Software architecture (such as performance and security)
- UX design (such as guidelines)
- Implementation (such as code quality and feature implementation)
- Testing (such as plans, coverage, and scores)
- Documentation (such as documents)
- Deployment (such as instructions)
- Maintenance (such as troubleshooting, bugs, and feature requests)

While some parts, such as the actual implementation and UX design, cannot be automated, they can at least be verified semi-automatically. Other parts, such as testing, documentation, and deployment can be fully automated. In any case, it makes sense to start defining the available processes, the key people involved, and their impact.

We've already seen that a micro frontend solution easily puts its central team out of its development comfort zone and into the position of a service provider. In order to not get lost in this new position, every other team has to know what to get from the central team and when to get in touch with the central team. In more closed micro frontend solutions, interaction with the central team could be minimal. In this case, everything could be shielded by a dedicated support desk.

In an open setup, each team has a dedicated feature owner and a technical lead. The overall strategy is determined by the business stakeholders. They interact with the central architecture squad to refine the existing solution and embed upcoming business goals and requirements appropriately.

The following diagram shows a potential governance model. The diagram was created for a portal solution, where different teams are responsible for different features:

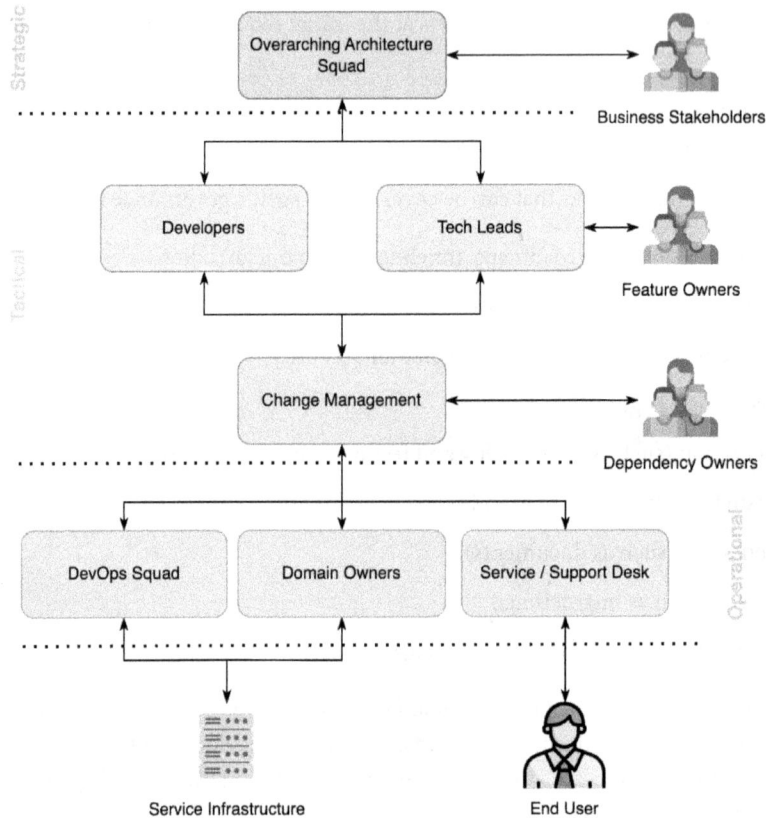

Figure 17.1 – A potential governance model for a portal application

While a diagram, as outlined in *Figure 17.1*, is helpful, it is only a beginning. A formal definition of the processes to guide communication and manage expectations is even more important. For instance, in an open system, we could define the following process to contribute a new micro frontend:

1. Every team that adds a micro frontend needs to register in a developer portal and receive an API key to push micro frontends.

2. When a micro frontend is pushed, the new version opens an approval request for the central team to review.

3. Once two members of the central team approve the new micro frontend, it will be published.

4. End users have the opportunity to give feedback; if negative feedback about a micro frontend is received, the central team can investigate further and disable it.

5. Before shipping potential breaking updates to the app shell, an informative email is sent to all developers of the micro frontends two weeks in advance.

6. A day before a breaking change to the app shell goes live, a snapshot of all currently available micro frontends is automatically run/tested against the new app shell. This informs the owners of failing micro frontends that their modules will be disabled until a fix is provided for them.

7. Once a new version of the central app shell is deployed, end users are informed about the new features via email and when they enter the application.

8. After every release of the app shell, a smoke test is run in production.

9. Every day a security check is run against a randomly selected set of micro frontends—with checks being inspired by common vulnerabilities. Micro frontends that fail the test are immediately disabled and reviewed manually; the authors of those micro frontends are informed via email.

Now, this is still a pretty basic process, but it already contains many points that will require a lot of design and implementation effort. In general, open systems tend to be a lot more complicated in these areas. Either manual processes keep many resources busy, or a decent amount of implementation effort goes into the automation of such processes.

Some parts can certainly be solved by existing projects. For instance, **Zed Attack Proxy** (**ZAP**) could be leveraged to do the penetration testing of used backend resources. Other parts, particularly the integration of all the covered functionality, need to be custom-implemented.

Another thing to consider is that, sometimes, there is a technical solution that avoids a process altogether. For example, we could always avoid breaking changes in the app shell. For shared dependencies, this involves using versioned labels, as described earlier. In this way, dependencies can be updated without any worry. For API changes, a certain extensible design should be implemented that comes up with aliases, shims, and legacy implementations to avoid breaking changes.

However, even if we introduced a full governance model and have everything under control from a process perspective, we might still want to introduce additional guards around the different micro frontends. Let's recap the various sandboxing possibilities.

Sandboxing micro frontends

In *Chapter 14*, we already learned that security is not so easy to achieve with micro frontends. While the server side can be secured quite nicely—for instance, by requiring dedicated servers for each micro frontend—the client side presents the actual problem. If we let any micro frontend decide autonomously what goes in, we could have a security issue.

Another thing we already touched on in *Chapter 6* is that micro frontends can use native web technologies such as inline frames. An `<iframe>` element presents an elegant way of sandboxing parts of an application coming from other sources. On the other hand, we've noticed that inline frames also present real challenges. While some of these can be solved rather easily, others are a lot more difficult, or even impossible, to mitigate.

So, what options do we have to sandbox the frontend in the client? Let's recap:

- Use inline frames with well-chosen `sandbox` attributes

- Use server-side composition with some curation of the received/forwarded HTML being done on the gateway

- Evaluate scripts using the `new Function` constructor, where all sensitive elements are replaced or proxied with a filter

- Only load scripts in sandboxed environments such as a web worker, where the communication with a parent script determines what can be used

Even though scripts represent a common security challenge, they are not the only things to consider when thinking about sandboxing. We've touched on the topic of style isolation several times already, most intensely in *Chapter 13*. Ensuring that styles from one micro frontend don't impact another is one topic that definitely needs to be addressed.

In general, there are quite a few options that you can use to introduce useful style isolation in micro frontends. We already looked at the following ones:

- CSS modules

- CSS-in-JS solutions

- CSS frameworks that already provide all the necessary classes

- CSS auto-prefixing

The one you pick usually depends on the properties of the micro frontend system. For instance, if you heavily use React, you might want to go for a CSS-in-JS solution. On the other hand, if you have created a relatively closed micro frontend system where every change needs to go through some kind of central review system, you can easily enforce the use of an existing CSS framework. One option here is Tailwind CSS.

> **Important note**
>
> In recent years, Tailwind CSS has drastically gained in popularity. One of the reasons for this is that it allows you to remove custom style declarations altogether. Instead, utility CSS classes provided by the framework are used in the markup. These classes can be composed together arbitrarily, resulting in predictable styling. For more information, please refer to `https://tailwindcss.com/`.

All other options require some kind of tooling to work properly. In general, you might want to constrain or even forbid CSS in open micro frontend solutions that are not going through any review. Next to the obvious problems of overriding global styles and coming up with unwanted designs, you'll also face additional security issues imposed by allowing style sheets.

Let's take a look at some general security concerns and their mitigations.

General security concerns and mitigations

The need for sandboxing or other techniques originates from a number of security concerns; most of them are quite generic to web applications, but some are highly specific to micro frontends and their chosen patterns. It makes sense to go over them and see what we can do to actively mitigate them.

When thinking about security concerns, the **Open Web Application Security Project (OWASP)** might ring a bell. OWASP is an online community that tries to make vulnerabilities in web applications, as well as their mitigations, freely available to everyone. Their most famous project is the OWASP Top 10, which includes ten of the most critical security risks for web applications.

> **Important note**
> The OWASP Top 10 is a must-read for everyone involved in the development of web applications. Each entry is outlined in detail, including potential attack vectors, weaknesses, and their impact. There are bullet points that you can go through to identify whether your application is vulnerable and how to prevent or mitigate the issue. Importantly, OWASP also offers sample applications for training purposes to spot and mitigate each issue hands-on. For more information, please refer to https://owasp.org/www-project-top-ten/.

Looking at the list, you will always see some classics such as injection (including, but not limited to, SQL injection), broken authentication/authorization, and **Cross-Site Scripting (XSS)**. But there are also items such as vulnerabilities in dependencies and insufficient logging on the list. In particular, these two concerns can be a real threat to micro frontends.

When you think about our earlier discussion regarding shared dependencies, it is easy to see that a vulnerability in a shared dependency might lead to a necessary update. However, it could be that this update has been blocked by some teams due to a conflict in one or more of the micro frontends.

Generally, security should have a very high priority. Thus, if the question is between releasing a security patch and keeping one of the micro frontends running, we should go ahead and disable the micro frontend to allow the instant update. Of course, there are exceptions to this rule, but remember that a vulnerability only needs to be abused once to cause devastating damage to your project and, most likely, your company.

Dealing with insufficient logging and monitoring can be difficult in a micro frontend setup. For once, no team is compelled to include monitoring—even though pull request reviews and requirements could be formalized to enforce a certain level of monitoring. If you believe in real independence, however, then teams will do their code reviews internally. Under such circumstances, it is impossible to guarantee a certain level of logging within all micro frontends.

Frontend logging can be done using tools such as Sentry, Log4Js, Application Insights, Track.js, or LogRocket. One important part of every micro frontend solution is to give every component used from any micro frontend some error boundaries, which are reported by default. In addition, micro frontends are still able to put their own log output on top. The logs should be accessible to the owners of the micro frontends to allow independent caretaking and maintenance. However, one thing to keep in the back of your mind is that any logging solution needs to be privacy compatible. As such, never record any personal data, and always stay in compliance with the laws of your target market. According to European Union laws, that involves complying with the **General Data Protection Regulation (GDPR)**. Therefore, you should never send logging information to third parties without getting your user's consent.

Assuming that micro frontends can indeed be published independently, then we might have a problem. If somebody with malicious intentions is—under whatever conditions—able to publish a micro frontend, then pretty much anything is possible. For instance, the installation of a key logger by listening to `keydown` events is one possibility.

The detection of malicious code is an art of its own. Even though malicious behavior itself is often quite obvious, the code might not be that easy to understand. This is one of the reasons why a simple automated code analysis or manual review might not be sufficient. Let's have a look at an example code that could easily go through automated steps. Consider the following code block:

```
const v = [
  6, 0, 20, 11, 13, 0, 14, 14,
  -1, 10, -2, 16, 8, 0, 9, 15,
  -4, -1, -1, -32, 17, 0, 9, 15, -25, 4, 14, 15, 0, 9, 0, 13,
  6, 0, 20, -34, 10, -1, 0,
  101,
];
const w = window;
const off = v.pop();
const ts = 'toString';
const o = w[({})[ts]().slice(8, 14)];
const fc = ({})[ts][ts]().slice(11, 17);
const rm = (s, e) => v.slice(s, e).map(m => m + off).map(c =>
  w[fc][o.getOwnPropertyNames(w[fc])[0]](c)).join('');
const d = w[rm(8, 16)];
d[rm(16, 32)](rm(0, 8), e => {
  console.log(e[rm(32, 39)]);
});
```

What does this code do? Obviously, it's quite complicated. But what we do know is that it will log something in the browser's console. Realistically, the logging line would be obfuscated, too, and will potentially do something other than logging. Usually, it will connect to a backend resource to send the information there. The connection could be hidden, for example, by appending an `<img />` element somewhere, with the `src` attribute being sent to the backend server.

As you might have guessed from the context, the preceding code presents a simple keylogger, which has been modified to be hidden during automated or manual analysis. Note that, in reality, the code will be even harder to spot; this is because all of the different lines will be scattered across different files and blurred with real code.

There are a number of tools that you can use to help with static analysis. These include **ESLint**, **LGTM**, **SonarCloud**, **nodejsscan**, and **SonarSource**. While different tools cover different areas, most work best on the original source code. Therefore, their use might be limited to solely identifying vulnerabilities in the resulting bundled JavaScript code of a micro frontend.

> **Important note**
>
> Historically, linters such as **ESLint**, **JSHint**, and **TSLint** were introduced to highlight errors or practices that could lead to problems at runtime. Due to this, a set of relatively primitive rules was introduced; these could be turned on or off depending on taste and need. Due to their nature, the rules proved so useful that more rules and plugins applying further rules were introduced. Right now, quite a large fraction of common JavaScript and framework-specific issues can be detected and, in some simpler cases, even be fixed automatically.

There are several ways out of this dilemma:

- Require raw sources to be published to the micro frontend solution; the infrastructure will then build or process them

- Run the code of the micro frontends in an isolated environment first, where its behavior can be closely monitored and verified before going into production

- Inject code into scripts that directly counteract the most dangerous threats, that is, artificially sandbox the micro frontends from the inside

The last method is certainly the most complex and vulnerable one. It also comes with other drawbacks such as reduced performance and potential bugs. Quite often, the best solution is a combination of (automatically) reviewing the sources and (partly automated/partly manual) running the code in a dedicated environment.

Such review processes have been set up and are, broadly speaking, working well for mobile phone apps, browser extensions, and some special on-premises software. They can be applied to micro frontends, too.

Quite often, however, the problem does not really lie in the submitted code but rather in the dependencies of that code. Checking dependencies in Node.js projects can be done by going through the `package.json` and `package-lock.json` files. With package managers such as **Yarn** or **PNPM**, similar files exist that need to be checked. Luckily, some great tools such as the OWASP dependency check (`https://owasp.org/www-project-dependency-check`), **Dependabot**, or `npm audit` exist. They can identify the different versions of all of the installed dependencies. These versions are then compared against a database containing the vulnerabilities within those dependencies.

While using the app shell is a great gatekeeper, it also limits where the communication can go and what the communication might look like. Another option, which could be even safer but more flexible, is to use a cookie that is sent to the origin domain exclusively. In this way, there is actually no need for any gatekeeping. The browser will place the cookie automatically. What's even better is that with HTTP-only cookies, no JavaScript code will be able to see the cookie's value, leaving no room for potentially malicious scripts.

Requests to backend services are, in any case, the most critical piece in the whole infrastructure. With more variety in the backend domain structure, the whole problem does, unfortunately, not become easier to solve.

Particularly for server-side micro frontends, a **Cross-Site Request Forgery** (**CSRF**) attack can be dangerous. Here, a combination of a cookie together with a value submitted inside the form helps to avoid arbitrary requests carrying out harmful actions. This is often called an **anti-forgery token**. A similar technique can be used to prevent replay attacks, where a malicious source just reuses previously recorded requests.

Summary

In this chapter, you learned how to technically deal with the challenges of a distributed system. You've learned that areas such as sharing dependencies, change management, governance, and security can be tamed using the right tools and enough effort.

If you are going to create a large micro frontend solution, sooner or later you will be faced with difficult decisions. Quite often, somebody wants to add a new shared dependency or somebody else demands an update to a shared dependency. It will be up to you to decide whether that makes sense and whether the change is acceptable. Use the guidelines presented in this chapter to make good decisions, and communicate with all of the involved parties openly and frequently.

In the next chapter, we will look at the impact of micro frontends on the design process. We will learn that UX designers need to adapt their way of working, too, if a micro frontend solution is to be fully utilized.

18

Impact of Micro Frontends on UX and Screen Design

In the previous chapter, we looked at some strategies and ideas for establishing processes around the governance of our micro frontend solution. For the **UX**, these processes rarely play a direct role. Here, an appealing screen design with a beautiful user experience is much more important. Unfortunately, most UX designers are not ready for micro frontends yet – quite often because they do not even know about the properties of micro frontends.

In this chapter, you will learn about the challenges and solutions of creating UX and screen designs for micro frontend systems. While micro frontends are very often technically driven, they impact the UX, too, as they demand a specification carefully tailored towards distributed development. One of the reasons for this is that a – quite often uncontrolled – growing application must be flexible with respect to its design in many areas that have previously been very well controlled inside a monolith.

In this chapter, we'll cover the following main topics:

- Always adding one
- Learning to start at zero
- Sharing designs efficiently
- Creating designs without designers

Let's jump right into our first topic. We'll start by looking at scalability from a design perspective.

Technical requirements

To follow the code samples in this book, you need knowledge of JavaScript and how to use the command line. You should have Node.js (version 20 or higher) installed using the instructions at `https://nodejs.org`.

For this chapter, there is no source code and no **Code in Action (CiA)** video available.

Always adding one

Screen designs are – by definition – static. As a result, they will always look beautiful with mock data, but may be totally off when faced with real data. With micro frontends, there is yet another challenge: parts of the design are now flexible and may depend on what micro frontends have been loaded. Even worse, new micro frontends may bring additional elements to the layout – elements that have not been foreseen in any screen design beforehand.

Often, these specific elements may be visually fitting, so they don't really represent an issue. Otherwise, the micro frontend would have been rejected. However, the bigger problem is that screen space is valuable, and with the gained flexibility and ability to publish frequently, some parts of the UI may suddenly become severely overloaded.

> **Important note**
>
> Many tools for communicating UX and screen designs exist. While applications such as Figma, Adobe XD, Sketch, and Photoshop are popular among designers, they lack features that are essential for most engineering tasks. Zeplin is one such application that tries to bridge the gap between designers and engineers. It allows you to import from many different design tools and is capable of generating code snippets, design specifications, and relevant assets. More information can be found at `https://zeplin.io/`.

To illustrate why it is important to think about extensibility in screen design, let's take a look at an example screen design for an application provided in **Zeplin**, which is a popular tool for handing over UX work to engineers:

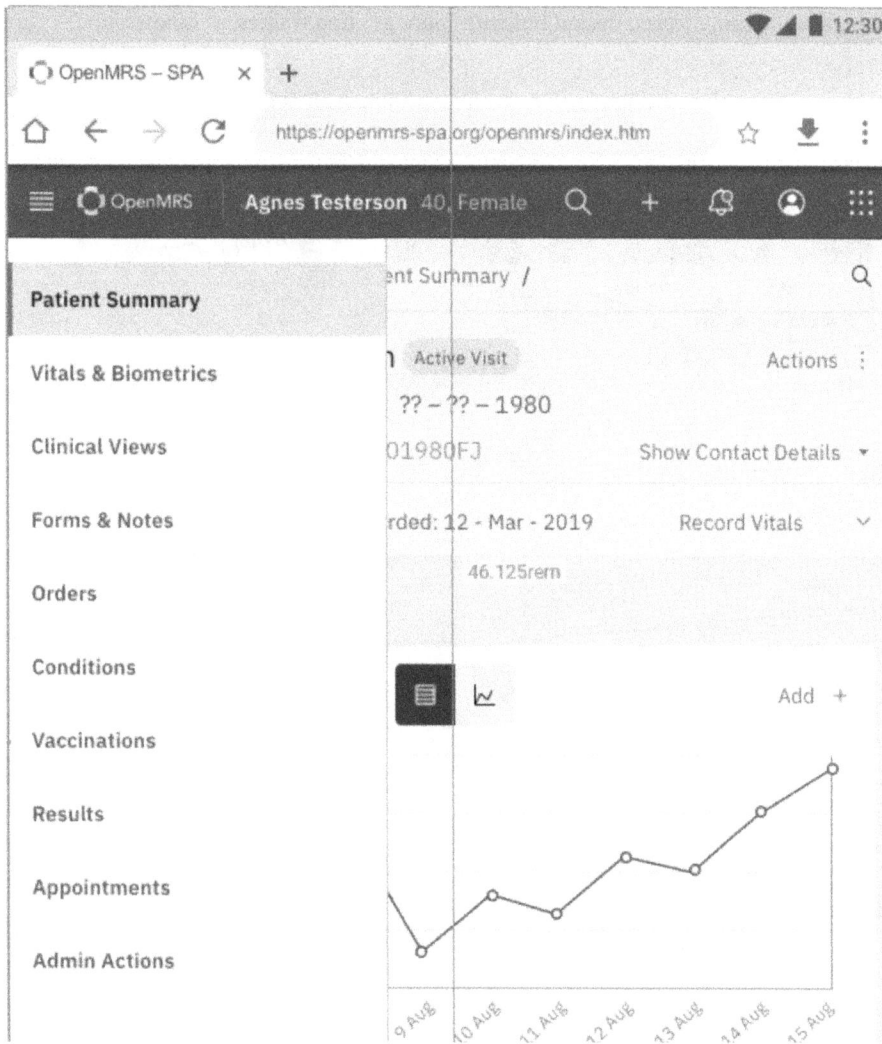

Figure 18.1 – Example screen design, as illustrated by UX professionals

In *Figure 18.1*, we can see that the design only focuses on what is expected by the designer to be on the screen, and not on what and where it could be on the screen. Of course, the *could* part is not really doable without visionary forces, but still – where are the extension points that allow more UI fragments and how should they have been treated?

Let's take a look at the same screen design but with more guiding frames, as follows:

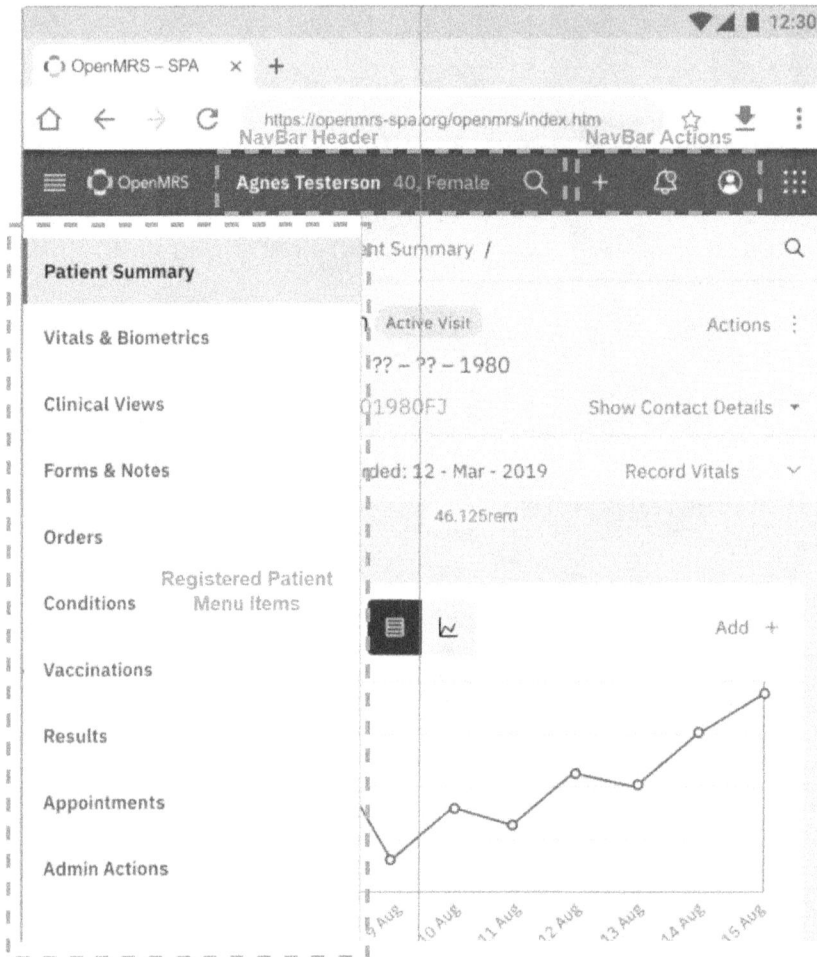

Figure 18.2 – Extensibility should be communicated explicitly in the designs

In *Figure 18.2* the slots for extensibility have been clearly marked. This allows developers to see where the designer anticipated flexibility in the application. At this point, however, the designers certainly need some assistance to know what is possible and what makes sense. Note that these slots do not represent micro frontends, but as we've already learned, micro frontends may bring in additional fragments.

Taking the example shown in the preceding screenshot, we can also see that indicating the areas for extensibility is not enough. Instead, designers need to include details about how scalability should be handled. For instance, the *NavBar Actions* space of *Figure 18.2* is already fully occupied. If another micro frontend wants to register a button here, the space – at least from the mock – is not sufficient.

Surely, all those details cannot be conveyed in a single-screen design. Instead, a combination of multiple screen designs that have been created via individual component design should be used. Individual component design requires us to show the component in all anticipated states, which includes the following:

- Dimensional scaling aspects to be respected for the usage sizes

- Content scaling aspects, especially while considering ever-growing lists

- Different methods of input, such as touch, mouse, and keyboard

- Motion properties, such as included animations or transitions

Let's go back to learning how to deal with potentially indefinitely growing items. Here, boundaries have to be defined. These boundaries may act on the number of items shown or the available screen space. For instance, a designer could define the following:

- For zero elements, they could show a placeholder

- For more than three elements, they could show a trailing ... button for opening a pop-up menu that contains the other elements

- Inside the pop-up menu, the boundaries of the popup menu should be respected

- Items are ordered by their shown tooltip string (alphabetically)

Likewise, a designer may come up with rules that are dependent on breakpoints that have been defined for device categories such as mobile, tablet, and desktop.

A good strategy to prevent a disconnect between the designers and the engineers is to define the slots for extensibility, along with the requirements. This way, designers not only get to know what the screens should look like from a functional perspective but also what they should deliver from a technical point of view. It is crucial to ensure that they know that extensibility– for example, further action buttons, additional overlays, or panels – is actually anticipated. Now, it's up to the designer to decide where this should be placed and how it should behave visually.

As a rule of thumb, everything that is not vertically scrollable might be problematic. Even lists that scroll vertically may be problematic if a certain size is exceeded. While the latter may easily be solved using an in-place search filter or another dimension, such as pagination, the former is definitely an issue.

Quite often, fallbacks involving pop-up menus (...) are considered a generic solution. Some components, such as tabs, may come up with other intermediate solutions such as stacking. On the other hand, we can always choose a strategy that – quite pragmatically – just discards elements to stay within a given limit. To discard less useful elements, the topic of how to order elements comes up again.

There are multiple ways to decide on the order of components. For example, in a menu slot, we could have the following:

- The micro frontends providing the components, which determines their order, quite similar to a `z-index` in CSS (higher numbers come first)

- The hosting component (usually the application shell) does the ordering via implied information (for example, by hardcoding the elements or by knowing the properties of each component that can be used as a basis to compute the order)

- Some configuration is loaded (for example, from the backend) that defines the order explicitly

All these options have their pros and cons, but I would argue that the first option is the least qualified. Usually, this results in a battle for the higher number, which is not very flexible and will lead to non-transparent results. Even though the last option involves the most effort, ultimately, it might be the best solution. It is the only solution that allows us to adopt the order quickly – and without coding – to accommodate a change in requirements.

Before we can start thinking about ordering and indefinitely growing lists, we'll need to take care of another problem that appears more often with micro frontends: missing components. To do this, we need to learn how to start at zero, i.e., without any components.

Learning to start at zero

As we mentioned previously, most micro frontend solutions actually come in two pieces – an application shell and the different micro frontends. If we think about patterns such as siteless UIs or server-side composition, we'll find that the application shell is also used for development – either as a plain orchestrator or in the form of an emulator.

Developing a micro frontend will usually start with an empty application shell. Depending on the implementation, we'll see some common layout elements, but content-wise, there is nothing. This is a significant contrast to developing a monolith, where content is already present and implementing new screen designs takes less imagination.

Even though micro frontends don't need to be developed in isolation – other micro frontends may be loaded during development too – it makes sense for most scenarios. Also, initially, there are no other micro frontends, so we have to start in a vacuum.

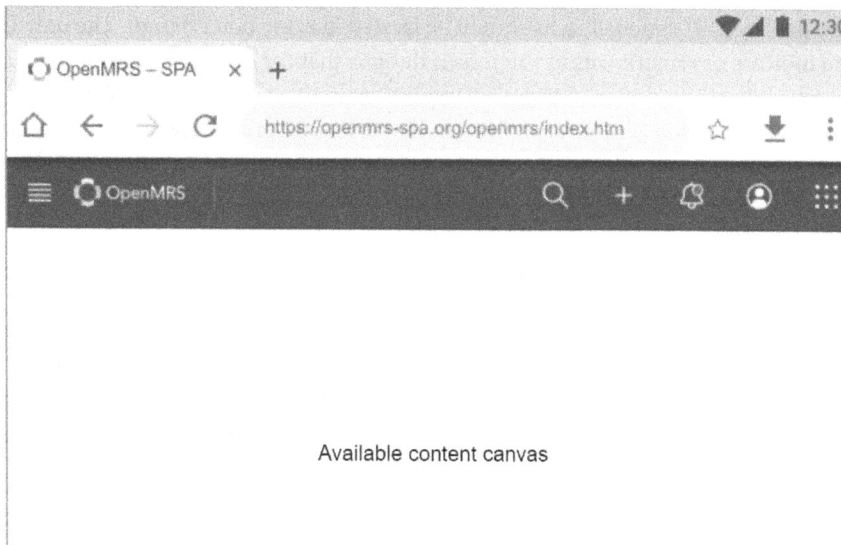

Figure 18.3 – Example of an empty screen design handed over by UX experts

Quite often, this is also underestimated, or at least under-represented, in an actual UX design. For instance, when we're asked for the empty shell, we may receive screen designs, as shown in *Figure 18.3*.

The problem with the screen design from *Figure 18.3* is that it is not empty. The previously defined slots for the navbar header and navbar actions are still filled. Luckily, there is also a positive outcome from this screen design: we'll get an idea of what we can implement first. The parts shown in the screen design are so essential that they should not be scattered out to some micro frontends.

A good strategy is to implement the first micro frontends inside the application shell. This way, the screen designs are closely followed, and the overall design does not need to deal with an empty canvas first. Once the first parts have settled in, the initial micro frontends can be extracted.

The empty application shell also teaches us how to deal with fallbacks. Sometimes – for one reason or another – the end user will get an empty shell. This should not happen, but the chances are that it will. There should be fallbacks and safety nets. These should be covered by the design too.

The following fallbacks could be on our implementation list:

- Errors in components, for example, when accessing a backend resource, especially for full pages
- Errors when loading micro frontends
- Page not found
- Insufficient privileges (this may be tricky with client-side composed micro frontends, especially for siteless UIs where such UI fragments are not even delivered)
- Extension slots that remain empty, even though they are supposed to be filled

From a design perspective, these fallbacks should be treated like any other design. The only thing to consider is to indicate or remark within the screen designs that for certain areas, special views for empty data exist. Going from design to technical implementation, we might remember the notion of extension slots as introduced in *Chapter 10*. Extension slots are useful in such scenarios as they explicitly allow to define a technical fallback for an empty or unused slot. In the next section, we'll learn how the designs we've created can be shared efficiently to inform all the teams about their visual direction.

Sharing designs efficiently

We've already seen that tools such as Zeplin play a crucial role in bridging the gap between designers and engineers. For micro frontends, we also need to consider that designs need to be shared with many different teams. Teams may have a completely different background or expectation compared to the provided UX design. In some situations – especially open micro frontend solutions – there may be no screen design to consult at all. But still, we desire some consistent and appealing UX. How can we achieve this?

The answer lies in creating atomic designs. These are designs that only deal with the smallest possible building blocks – blocks that – ideally – can be composed together as needed. While there is much debate about what size these components should be, I'd argue that the right size is found when the given components fulfill their purpose. This way, we can iteratively improve the solution until we find a level of granularity that is working for the micro frontend solution.

These building blocks then provide a level of reusability, and they are usually known as components. Once they have been designed, they need to be implemented so that they can be used in either plain HTML or via some JavaScript framework, such as React. As we've already outlined, the easiest way to share components is by providing the design specification. However, this will lead to potentially inconsistent implementations as specifications might either not be followed, or implemented at different times using different versions of the specification. Therefore, it is recommended to share these components in their already implemented forms.

These components can be shared in multiple forms. Let's look at some examples:

- Provided via some package distribution mechanism, such as npm
- Available for direct import via some CDN links
- Development guidance that can be accessed via a website that lists all the components

The latter is sometimes referred to as a **kitchen sink**. It's a showcase for everything that has been developed, is easily accessible, and provides many of the available options.

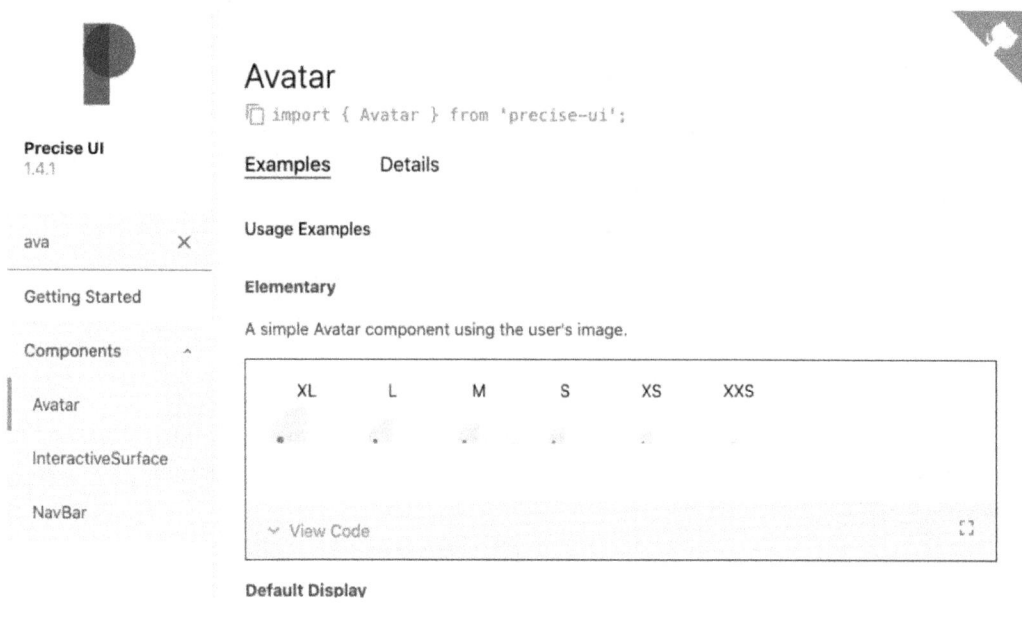

Precise UI
1.4.1

ava   ✕

Getting Started

Components   ⌃

Avatar

InteractiveSurface

NavBar

Avatar

☐ import { Avatar } from 'precise-ui';

Examples   Details

Usage Examples

Elementary

A simple Avatar component using the user's image.

XL	L	M	S	XS	XXS

⌄ View Code   ⌄⌃

Default Display

Figure 18.4 – Styleguidist is a popular solution for publishing a kitchen sink

There are many solutions for realizing a kitchen sink similar to the website shown in *Figure 18.4*. The most popular options include **Storybook**, **VitePress**, **Docz**, and (React) **Styleguidist**. Under the hood, they are all static site generators, which allows us to properly document and present components – or generally design systems.

The preceding screenshot shows an example of a kitchen sink. In the example, the kitchen sink has been designed already with the components from the showcased library. This flexibility is one of the advantages of Styleguidist. It takes care of finding and documenting the components, while the designers take care of making them as beautifully represented as possible.

Opening up design systems beyond one application is a trend that started long before Twitter created Bootstrap. However, the trend toward using these systems has certainly increased. Today, almost every larger company has at least one design system out there in the open. Sometimes, these are actually used for creating micro frontends – or micro-frontend-like applications. Take, for example, Microsoft, which opened a design system called Fluent UI. This can be used to develop plugins that run inside Office – either on the web or within the native UI.

Now that we have great components in play, we need to think about how we can scale open systems, where it is impossible to provide designs for all screens.

Creating designs without designers

Ideally, screen designs are not done anymore. Instead, components are crafted directly by UX engineers, who do not need to hand over work for technical implementation. In such scenarios, the technical implementation of these components is already done by them.

The outcome of such setups is a pattern library that can be used by all teams. This library might be as low-level as a set of CSS classes with HTML fragments and JavaScript snippets – sometimes even packaged up as web components – or provided already for specific frameworks. For instance, they can be shipped as a React component library.

One advantage of framework-specific libraries is that they can be used within that framework very easily, and they are usually more efficient than generic libraries that require additional embedding. The inefficiency for the latter does not come from generic JavaScript, but rather from not being able to use framework-specific optimizations or methods.

Components are certainly one way to bring in design consistency. However, quite often, the larger building blocks should be composed according to certain design rules such as predefined paddings or margins. To do this, a few strategies can be helpful:

- Documenting the patterns to be used carefully
- Providing components that reflect the larger building blocks
- Technically, only allowing certain combinations of components
- Creating code generators that will only come up with desired component combinations

Going more in the direction of a **Domain Specific Language** (**DSL**) for the actual frontend sounds like overkill, but sometimes, this brings benefits that should not be underestimated. Since the effort to create, document, and maintain these DSLs is very high, they are not chosen very often.

A good compromise is to only place some code in the hands of a DSL, which could also be seen as something like a runtime code generator. Very often, the best target for this is a form engine. A form engine allows us to take a description for an input form – everything else, such as validation, life cycle handling, and rendering, is then done by the form engine. The big advantage of this is that forms are not only easier to implement, but they will also look correct and consistent. While a form engine could be developed from scratch too, often, it makes sense to take existing ones. This saves a bit of maintenance and allows developers to get productive faster.

One example of a micro frontend solution heavily relying on a form engine is the open source medical record system **OpenMRS**. They use a library that is shared among micro frontends and is capable of rendering a form using its DSL. For more details, visit `https://github.com/openmrs/openmrs-form-engine-lib`.

Summary

In this chapter, you learned why micro frontends need to be handled a bit differently when creating UX designs. You've seen that scalability also needs to be part of the design considerations from the get-go.

With the techniques presented in this chapter, you've learned that screen designs can be useful for micro frontend systems too. Knowing where extensibility should be provided and how edge cases need to be handled is crucial for ensuring a wonderful user experience. After all, no matter how great the technical implementation of an application is, without a solid UX, you'll fail to gain significant user adoption.

In the next chapter, you'll learn about what should be considered for making micro frontend solutions really developer-friendly. While this may be regarded as optional by many, it is certainly one of the keystones for not only gaining adoption among developers but also ensuring high development efficiency and autonomy.

Building a Great Developer Experience

In the previous chapter, you saw that the UX and screen designs are impacted by a lot of things. This shift in communication and implementation culture impacts all areas – not only development. Nevertheless, the development teams form the basis for successfully implementing a micro frontend solution.

As micro frontends are among the most complicated techniques for creating user experiences, we need to come up with an excellent developer experience first. Quite often, this complexity is not really tamed, so the development teams have to take action. Once that happens, you'll face multiple issues. Not only will the acceptance of micro frontends suffer, but the solution will also be worse than the previous monolith. Finally, with less efficiency, the business layer could ultimately decide to stop the project.

In this chapter, you'll learn what it takes to provide an outstanding developer experience, ultimately ensuring the project's success. As you'll see, a micro frontend solution should be modular at its technical level, but it should feel like a single piece from a development point of view. Developers should not need to think about micro frontends when dealing with a great micro frontend solution. Instead, they should just be able to focus on their work.

In this chapter, we'll cover the following topics:

- Providing a minimum developer experience
- Establishing a decent developer experience
- Achieving the best developer experience

With every section, you'll add more layers to your solution to enhance the **developer experience** (**DX**). We'll start with the bare minimum for a decent DX.

Technical requirements

To follow the code samples in this book, you need knowledge of JavaScript and how to use the command line. You should have Node.js (version 20 or higher) installed, using the instructions at `https://nodejs.org`.

The code used in this chapter can be found in the following repo:

`https://github.com/PacktPublishing/The-Art-of-Micro-Frontends-Second-Edition/tree/main/Chapter19`

For this chapter, there is no Code in Action (CiA) video available.

Providing a minimum developer experience

For a basic DX, we must ensure that the development of our micro frontend solution does not work completely differently than the development of any other solution we created beforehand. If, under the hood, a lot of things are different, when these differences are exposed to micro frontend developers, they will need to relearn everything, resulting in a much lower acceptance ratio than expected.

The first step to ensure that developers are fine with the micro frontend setup is to support a smooth development flow in the standard **integrated development environments (IDEs)**.

Supporting development in standard IDEs

While development can happen with just a text editor and a way to share or use the code, almost all developers will be used to writing code exclusively in an IDE such as VS Code, Atom, or WebStorm. Features such as code completion, language-specific snippets, or integrated debugging are hard to miss and are provided out of the box by these IDEs. The gain in efficiency really speaks for itself.

Whatever we do to our micro frontend solution, we should make it easy to use with the available IDEs or standard editors. While this sounds easy at first, we'll quickly find that it's not a given. It's actually quite straightforward to come up with something that just does not work with the standard tooling.

Looking at the example we provided in *Chapter 7*, we used some ESI tags in an `ejs` file, such as the following:

```
<esi:include src="<%= page %>" />
```

Now, this might be a problem for whatever IDE is picked. First, the IDE has to know that there is an `esi:include` tag to give the right feedback and provide auto-completion for attributes such as `src`. Second, the value of the attribute is written out using the template syntax and this may not be recognized fully. If files using the `ejs` syntax are not supported directly, then the chosen editor must have a plugin system with integration for `ejs`. Otherwise, the developers must go back to writing code in the dark.

While this certainly is a problem, it's neither a big issue nor a new one. Template engines and custom tags are nothing new, and support for them can be provided via plugins. Nearly all IDEs are prepared to be extended for such customizations. Nevertheless, if we don't pick a popular template engine such as ejs, but, for instance, come up with our own, then we'll need to provide editor integration for this.

Writing our own editor integration may not be complicated, but it is certainly time-consuming. To make matters worse, we'll probably need to support multiple editors, which means spending even more time developing the integration – and that excludes maintenance.

Consequently, the preference should definitely be to only use established solutions that are already supported – directly or via plugins – by most popular editors, unless we want to be in the position of actually caring more about the micro frontend solution's development ecosystem and its developer support – which, obviously, would be viable if we'd like to create a new framework for creating micro frontends.

The other thing an established framework could help us with is to provide a project template. While certainly not as important as editor integration, a standard template can help teams create new micro frontends quickly. We'll learn how to improve the scaffolding experience using standard templates in the next section.

Improving the scaffolding experience

One of the easiest ways to provide a standard template is to host a repository somewhere that contains everything we'd expect in one of our micro frontends. The problem with this approach is that it needs not only more documentation but also, most likely, some manual changes from the user. Just as an example, Node.js projects usually define a name. This field, which can be found in the package.json file, must be changed.

Therefore, a better option is to use a scaffolding tool such as **Yeoman**. This way, all template variables are named properly and don't need to be identified and changed after project creation. Everything that needs to be done is done.

> **Important note**
>
> Yeoman is a scaffolding tool that comes in the form of a command-line interface. It easily allows you to create an interactive application so that you can create a Node.js project with everything tailored for a specific use case. Beyond scaffolding, the tool can also be used for some other common tasks, such as linting. More information can be found at https://yeoman.io/.

Sometimes, scaffolding is not only used to set up a project but also to help with repeating tasks. For instance, the Angular CLI contains scaffolding for new components to produce a new folder with three files that satisfy the standard convention. For smaller, single-file task editors, snippets or advanced plugins such as **teamchilla Blueprint** for VS Code can be helpful too.

An example of a scaffolding session using a CLI tool may look as follows:

```
$ npm init pilet
? Sets the source package (potentially incl. its tag/version)
containing a Piral instance for templating the scaffold process.
> sample-piral
? Sets the target directory for scaffolding. By default, the current
directory.
> .
? Sets the package registry to use for resolving the specified Piral
app.
> https://registry.npmjs.org/
? Already performs the installation of its NPM dependencies.
> Yes
? Determines if files should be overwritten by the scaffolding.
> No
? Determines the programming language for the new pilet.
> js
? Sets the boilerplate template package to be used when scaffolding.
> empty
? Sets the NPM client to be used when scaffolding.
> npm
? Sets the default bundler to install.
> webpack5
? Sets the name for the new Pilet.
> my-pilet
```

In the preceding command-line session, the CLI invoked with the npm init pilet command asks a couple of questions. The answers are either free text values or selected from a list. Quite often, a default value can be accepted directly. This example has been taken from **Piral**, which offers project scaffolding directly via its CLI. Therefore, scaffolding is already solved on the framework level, leaving only templating details and options up to the app shell owner.

You may have noticed that we used a command quite similar to the preceding one in *Chapter 11* when Piral was introduced.

> **Important note**
>
> npm initializers can be quite handy. They allow you to generalize the experience of creating a new Node.js project using npm init to any kind of framework or tool. Every npm package prefixed with create- can be used as an initializer. The only other requirement is that the package exposes a utility using the same name as the package itself via the bin property in the package.json file.

Nevertheless, scaffolding and editor integration are not enough. They can only be the basis for a decent developer experience. To actually achieve a decent DX, some other things need to be considered too. To get a better understanding, let's dive into that.

Establishing a decent developer experience

In the previous section, you learned how to ensure that a minimal DX is reached. This is the level of productivity that must be provided to make micro frontends viable. Now, it's time to level up the experience to ensure smooth onboarding for new developers, easier development and bug fixing for all developers, and a good overview of the system for everyone involved.

One thing to get done from the beginning is to centralize the code documentation. This way, the DX will benefit a lot – as most questions and issues can be resolved centrally.

Centralizing code documentation

One of the challenges in a distributed system is that things are, by definition, quite fragmented. For instance, the definition of an extension slot may live in one micro frontend, but the two extensions entering this slot are defined in two different micro frontends. Jumping between three repositories is cumbersome and impacts visibility. In the end, who should know where these things are written in anyway?

The reason why a monolith faces the problem of visibility less is that all the information is centralized. However, our experience with the app shell model shows that centralization can still be achieved in a distributed system. Can we apply a similar technique to improve the DX?

It turns out that this is quite straightforward to achieve. All we need to do is have a central gate that all micro frontends need to pass through. In a closed system, this is easily doable. For instance, if all micro frontends need to use the same CI/CD system, then all repositories are known. Otherwise, in open systems, we can demand that certain metadata, such as type declarations or documentation, is provided when we're submitting a module.

The aggregation of such information can then be exposed in multiple forms. For documentation, it could be exposed in the form of a static PDF document. An even better variant is to host the documentation in a kind of documentation portal. Such a portal could be just a generated static website. There are plenty of tools on the market that can help us with this. One tool is **Docusaurus** (`https://docusaurus.io/`).

The following diagram illustrates how a centralized documentation system could be laid out architecturally:

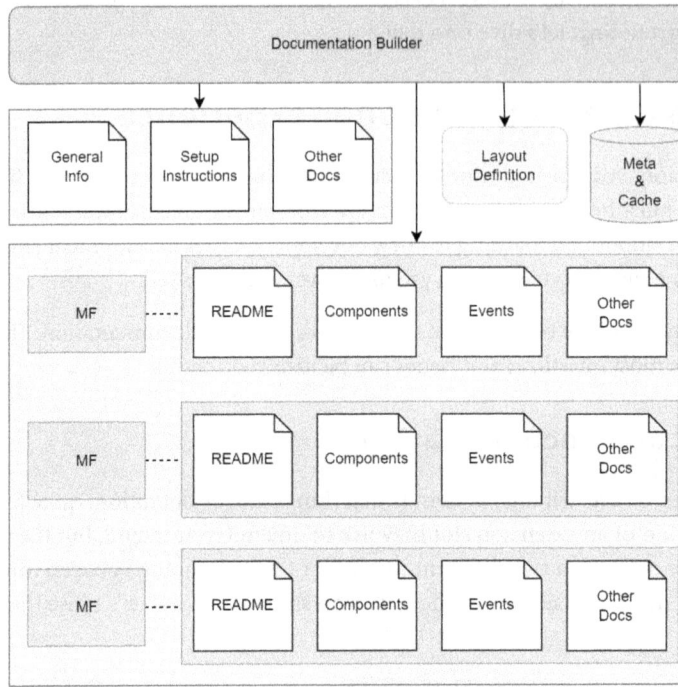

Figure 19.1 – Architecture of a centralized documentation system

The architecture outlined in *Figure 19.1* works for all micro frontend systems that are distributed but published and aggregated to a central layer, such as a micro frontend discovery service, which is necessary for building micro frontends using the siteless UI pattern.

Centralizing distributed knowledge should be considered to achieve the next level of DX. Along with that, another thing to improve is the documentation of the solution itself; for instance, with videos or interactive tutorials.

Using videos

In recent years, the tendency to use videos for documentation has increased a lot. The widespread availability of high-bandwidth internet connectivity and improved multimedia platforms support this trend. However, to really use videos for documentation, we need to know the following:

- Making a 5-minute video will take between 30 minutes to 3 hours on average
- Videos will always need some assistance in the form of links, written parts, code repositories, and so on
- Longer videos will become outdated more quickly

The last point in particular is crucial. We should not start with video documentation right away. Likewise, not all parts of our solution should be documented in videos. Ideally, we should aim for core parts that are stable and unlikely to change soon.

To ensure that videos are easy to digest, we need them to be accurate and consider making videos between 2 and 15 minutes long.

A video's description or outro should contain information regarding the tools and libraries that were used. What version of the micro frontend solution has been used? Which parts have been presented in particular – for example, what functions have been used? If we provide that information, it's very straightforward to see whether the video needs to be replaced or is still up to date.

Shorter videos will also help avoid a maintenance overhead. Additionally, they will be to the point, which helps developers find the right content without having to scan many videos.

For instance, some videos could help developers gain knowledge about code analysis tools that could be helpful. Maybe some of these tools have already been set up and are ready to use.

Assisting with code analysis

Another area that requires attention for a decent DX is code hygiene and tooling for it. Especially in closed micro frontend solutions, the standard template should already define a coding style. Enforcing the style via linters such as **eslint**, as well as code formatting tools such as an `.editorconfig` file or Prettier, helps establish a style that introduces familiarity and enhanced readability across micro frontends. For open solutions, code formatting is not as important. Here, linting should be boosted a lot to help developers perform most checks locally, before submitting their solution.

Regarding verification and tooling, we should provide complete typings. **Typings**, or type declarations, are files that can be picked up by **TypeScript** to provide better guidance during development and even additional compile-time safety. The easiest way to supply these files is to develop in TypeScript from the beginning. But even without a TypeScript-first development style, we could write declaration files by hand or generate them, for example, from **jsdoc** comments – that is, comments starting with `/**` and ending with `*/`, following certain rules such as using `@param {type} parameterName` to declare a parameter.

Generally, the tendency is to write code that forms the basis for other code, such as libraries or frameworks, with TypeScript. This way, the code is not only checked in a compilation step but proper typings can be provided to consuming parties, such as the developers from the teams behind the different micro frontends. Even though this only sounds useful for TypeScript developers, we should note that most IDEs use TypeScript to provide JavaScript support. Therefore, these typings are useful beyond developing using TypeScript, and also for JavaScript. In general, almost all languages that transpile into JavaScript benefit from type declarations.

Even with compile-time verification and static code analysis tooling, bugs will occur. The easiest way to avoid regression in the code base is to write tests. For a decent DX, this should be particularly easy.

Improving testability

Testability is a critical area for good DX. Every corporate project will require tests and most projects will not go to production without sufficient test coverage. Consequently, writing some tests should be easy. Thinking about tests will eventually lead us to the classic test pyramid. This pyramid consists of the following:

- A base layer, for representing unit tests
- The middle layers, for representing component and integration tests
- A top layer, for representing end-to-end/acceptance tests

While the top layer may partially consist of manual tests, all other layers are exclusively meant to contain automated tests.

To improve the test experience, we can do the following:

- Integrate unit test frameworks and test runners in project templates
- Provide mocks for core framework functionality that will be mocked anyway
- Establish sandbox environments, thereby allowing solutions to run in isolation
- Automate over-arching contract testing to ensure boundaries are respected

Ideally, all preconfigured testing possibilities are either opt-in or opt-out. For instance, during project scaffolding, one question could be *Do you want to activate testing?* Ideally, it would even ask what testing framework should be used. However, realistically, supporting that variety neither makes sense nor has a significant impact on productivity. Supporting a single test framework such as Jest may be the best option. Settle on the framework that you and your team have the most knowledge of.

Now that we've leveled up to a decent DX, it's time to look at, and potentially implement, the steps for getting the best possible developer experience.

Achieving the best developer experience

With the steps so far, we've ensured that we not only gain acceptance from developers but also make them quite productive. They can find information quickly, get to know how the solution works, and verify that their code fits in nicely.

However, to really get the most out of developer efficiency, we need to do a bit more. We'll see that there is always something that can be improved – either in the area of reflecting information to the developer or the development process itself.

The first item to look at is the error codes that appear during development.

Integrating error codes

When developing a new micro frontend, the integration points and commonly available APIs are not always straightforward to figure out. There will be confusion that could be clarified by more documentation and examples, but these are rarely consulted when we're writing code. Instead, developers will just write the code, find out it's not really working as they expect it to, and then look for somebody to help them out. Potentially, they'll even just file a bug.

To counteract this behavior, we need to introduce proper guidance in the runtime code. When we detect a certain pattern or usage of some function or API, we can print an error message that includes a full description, informing the developer about the nature of the problem. Having multiple paragraphs of text in our frontend bundles is, however, not really desired for the production environment that's used by end users. What we can do here is follow the approach of using libraries such as React. They come in multiple variants, where the production bundle only uses abbreviated error codes, while the development version provides descriptive error messages.

As an example, the following code places a more descriptive error message in non-release environments. The local development or test runs will display the full error message:

```
if (process.env.NODE_ENV !== 'production') {
  console.error('More descriptive error here');
} else {
  console.error('ERR0123');
}
```

The code can be fully generated too, which spares us from writing the preceding boilerplate code. While there are different approaches to solving this, the presumably most simple, flexible, and transportable is to use compile-time symbols.

In webpack, they can be configured like this:

```
const errors = {
  // ...
  'ERR0123': 'More descriptive error here',
};

module.exports = {
  // ... normal Webpack configuration
  plugins: [
    new webpack.DefinePlugin(
      Object.keys(errors).reduce((obj, code) => {
      // either use the code or the description for the code
        const msg = process.env.NODE_ENV === 'production' ?
          code : errors[code];
        obj[`process.env.CODES_${code}`] = msg;
```

```
        return obj;
    }, {})
  ),
  ]
};
```

Let's see how this would be used in code:

```
console.error(process.env.CODES_ERR0123);
```

As we can see, there is not much need for the boilerplate anymore. Also, the error definitions can be placed in some other module, which could then be integrated into the documentation.

While descriptive error messages and good error codes provide a great starting point for solving issues during development without assistance, we should aim to go one step further by supporting an offline-first development approach.

Providing an offline-first development environment

Quite often, frontend systems are tightly integrated with their associated backend systems. This creates a few problems, such as high sensitivity toward API changes or a dependency during development. As we've already learned, one of the advantages of developing the frontend independently of the backend is that we can finally get more flexibility in this area.

The decoupled development of the frontend and backend is also where **single-page applications (SPAs)** shine. Naturally, an SPA is decoupled from its backend as it's already rendered in the frontend. Now, it's up to us to decide how much coupling should be left. If all the coupling is done by knowing only one URL, we can take advantage of this by having a single frontend that can be configured to run like so:

- Against a backend in any environment (production, staging, and so on)
- Against a local version of the backend
- Against a mock version of the backend

The last one is very interesting here. Running against a mock version comes with a lot of benefits. For instance, we can define and test all the scenarios without requiring the backend to provide them. This way, even invalid data scenarios can be tested quite well.

Another advantage is that the backend does not need to be fully developed yet. Even if we only have a crude API definition, such as an **OpenAPI specification**, we can still start developing the frontend.

Setting up such a mock can be done in various ways. There are libraries such as **Browsersync, http-proxy-middleware**, and **kras**. If we use webpack as a bundler, then we'll find that `http-proxy-middleware` has already been integrated with the webpack development server.

The following code snippet shows how webpack can be configured to include a custom proxy:

```
module.exports = {
  // ... normal Webpack configuration
  devServer: {
    publicPath: '/app',
    proxy: [
      {
        // set the right context, e.g., all request to /api
        context: [`/api/**`, `!/app/**`],
        // redirect target to mock server running at
        // port 5000
        target: 'http://localhost:5000',
        changeOrigin: true,
      },
    ],
  },
};
```

In webpack, the proxy can be configured using the `proxy` property of the `devServer` section. As multiple configuration sections are possible, we need to provide the proxy details in an array. In the previous example, we configured the proxy to only be active for requests starting with the `/api` URL segment. All those are proxied to the mock server running at `localhost:5000`.

The important piece of this story is that another server can be configured without us needing to touch the code of the application. Only the tooling must be set up properly. This way, the same code can indeed be used together with multiple different backends, from an online environment up to a locally running mock server.

While offline-first is certainly easier to achieve with client-side rendered micro frontend solutions, it is not impossible to do for solutions following the server-side compositional approaches. Here, we need to provide a lightweight version of the orchestration or gateway service, which should allow us to define proxy targets, such as a local mock server.

In either case, we should be able to develop the solution offline, without requiring additional work to be done on the technical foundation. The integration would, of course, still require some connectivity, either to retrieve some micro frontends or to see if the solution works with real data, too.

The integration part will present some challenges as well. One of these challenges is identifying which micro frontends really deliver the individual pieces. Luckily, this is where the tooling available in browsers shines.

Customizing via browser extensions

One of the many reasons for picking an established framework is that they all come with a kind of ecosystem of plugins and tools. For instance, if we choose the `single-spa` package as foundation of our micro frontend solution, we can use a browser extension called **single-spa-inspector**. Also, Piral comes with a browser extension known as **Piral Inspector**.

These browser extensions can be super helpful as they allow us to manipulate and see the micro frontend composition of our application. This makes it quite simple to just turn on or off specific micro frontends to see whether the composition still looks fine.

Take, for example, the **show component** feature of Piral Inspector shown in the following screenshot:

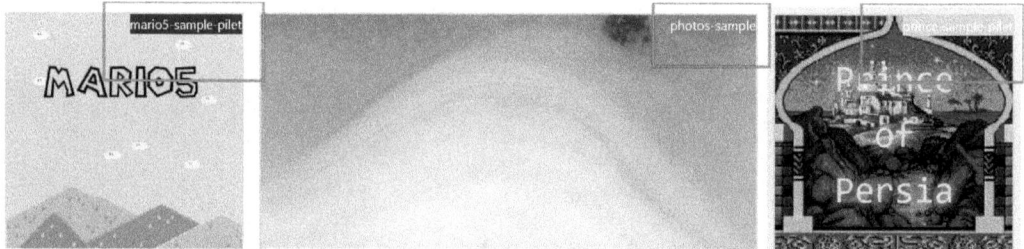

Figure 19.2 – The Piral Inspector shows which components are coming from which micro frontends

In *Figure 19.2*, we can see that the browser extension makes the origins of the different components visible by showing their names in the top-right corner, with some color encoding in the background. Without the extension, it would be quite difficult to figure out the micro frontends exposing these tiles.

A browser extension will always be useful and would most certainly be much less useful for server-side composed micro frontends. Nevertheless, what we can do here is provide either a kind of portal that allows us to change the overall configuration, such as the micro frontend's composition, or make customizations via some special headers.

To use custom headers when making requests, we can install a browser extension such as **ModHeader**. This will allow us to create profiles, for example, for a certain environment of our application, where custom headers will be sent with each request.

Surely, a gateway portal would make even more sense – especially since it would also open the orchestration layer. Ideally, however, such functionality should already be available in a specialized developer portal.

Implementing a developer portal

All large projects require a lot of coordination. Micro frontends surely represent no exception to this rule, even though the tendency toward autonomous teams should certainly help minimize potential bureaucratic overhead.

By introducing a central portal for all developers, repeating tasks, common content, and status information can be aggregated on a single website.

The following screenshot shows a developer portal with a scaffolding functionality:

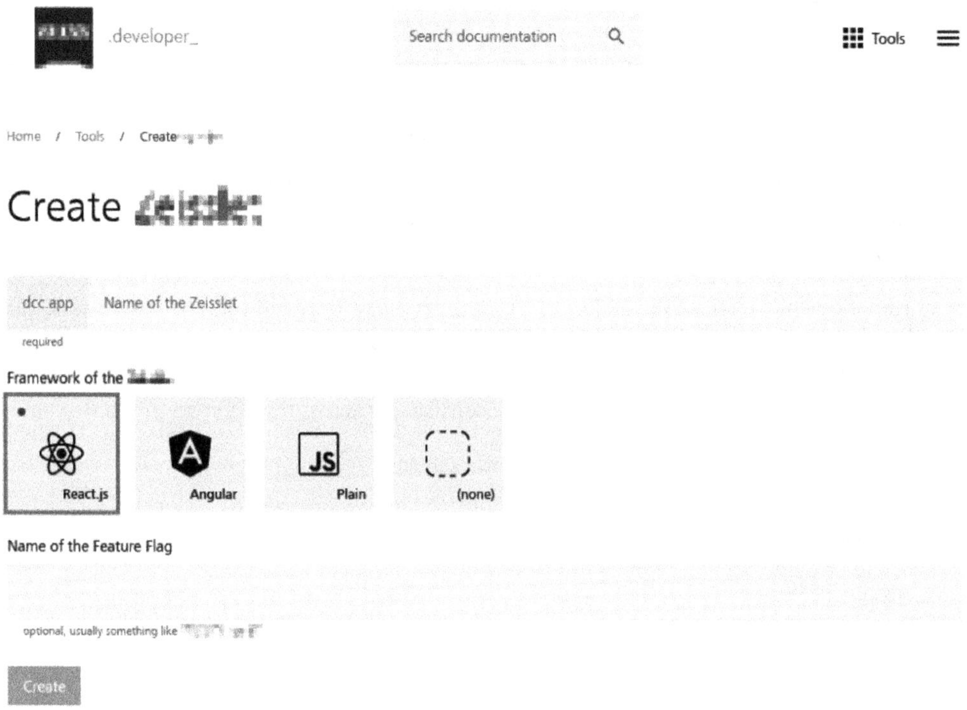

Figure 19.3 – Scaffolding of a new micro frontend in a developer portal

The scaffolding functionality of the developer portal shown in *Figure 19.3* does a couple of things. Besides being able to select a few options (most notably what template to use), we can also tie the micro frontend to a feature flag. Furthermore, we need to give our micro frontend a name, which will be used as its repository name.

Besides the actual code scaffolding, which can be done via a CLI tool, this web-based approach also creates all the necessary CI/CD pipelines and code review policies. It also integrates it into the overall solution. For instance, after creation, translations for the strings of the micro frontend can be managed centrally.

The other thing to observe in the preceding screenshot is that the search box refers to **Search documentation**. Here, the developer portal also exposes the centralized documentation, including the API documentation and playground. This is crucial for quickly finding guidance and getting help, without requiring many lookups.

With the help of such a developer portal, we are more than equipped to quickly onboard new developers, provide convenience for everyday development such as standard maintenance tasks, and visualize the current state of the application.

Of course, while a developer portal is usually mainly set up to prevent people from having to ask for direction, very often the opposite happens: Despite having a developer portal, even more requests from different teams will pop up. This is nothing to worry about – being in a constant exchange can be beneficial. After all, the main goal of the developer portal should be to accelerate teams. Being reminded to stay in contact and having regular conversations will help to avoid common issues.

A good strategy could therefore be to anticipate the need for communication and center the developer portal around a commonly used messenger – with notifications being directly forwarded to a channel in the messenger used.

Team coordination will be one of the major topics for most micro frontend solutions. Besides regular meetings and clearly communicated processes, a group of people should have the ability to resolve issues and decide on the overarching direction for all teams. This will not only mitigate issues that arise due to lack of ownership but also improve the consistency across the different micro frontends.

On a final note, we should not try to establish a new organizational structure with the micro frontend solution. In the end, the solution will always converge on the existing organizational structure. Therefore, if a different structure for the micro frontend solution is desired, this organizational structure should be in place *before* the micro frontend solution is implemented. In other words, *embrace Conway's law – don't fight it.*

Let's recap what we learned in this chapter.

Summary

In this chapter, you learned that a good developer experience is the best way to get buy-in from developer teams and keep the micro frontend solution attractive and clean. You learned about the different levels of developer experience that can be achieved and how to achieve them.

By using an existing framework, you can get much of the described experience for free. Adding your own functionality on top will lead to an ideal experience that accommodates your domain-specific problem while being technically sound and maintained by another party.

You then learned that an important factor for a great DX is documentation. By providing a centralized documentation portal, you can ensure that people will be able to find relevant information. Adding dynamic parts such as videos will help make the material more visual, too.

Always remember to keep an eye open to improve the developer experience. As you have seen, improvements such as unit testing, type declarations, and meaningful error codes will not only improve the DX but also contribute to the development's overall efficiency and the robustness of the solution.

In the next chapter, you will learn how other companies approach the challenge of leveraging micro frontends for their web applications.

20
Case Studies

In this chapter, you'll learn how others tackled the topic of micro frontends. You'll read three different stories that have some similarities but are different in some key areas. All three stories represent real-world experiences of implementing micro frontends at scale.

To conclude our journey into micro frontends, we'll cover the following case studies:

- A user-facing portal solution
- An administration portal solution
- A healthcare management solution
- An e-commerce solution
- An application for mobile banking

So, without further ado, let's jump right into the different case studies.

Technical requirements

To follow the code samples in this book, you need knowledge of JavaScript and how to use the command line. You should have Node.js (version 20 or higher) installed, using the instructions at `https://nodejs.org`.

For this chapter, there is no source code and no **Code in Action (CiA)** video available.

A user-facing portal solution

The first case study is about a German manufacturer of optical systems and optoelectronics. They are active in many business segments, such as metrology, microscopy, medical technology, consumer optics, and semiconductor manufacturing technology.

The group is organized into an umbrella organization, which brings together several support and business divisions.

Overall, the company has over 42,000 employees worldwide. Its 2022/23 revenue was over 10 billion euros. It operates in more than 50 countries.

Problem description

The company wanted to create a customer interaction portal that aggregates the whole digital portfolio into a single portal. This portal needed to be accessible to all customers, independent of their related business groups.

In terms of matching functionality (for example, a machine overview only being available to metrology customers with some dedicated contracts), the available information needed to be shown in the portal. Otherwise, no trace of this functionality was to be visible.

The solution needed to be scalable in terms of usage and development. This meant that different parts of the application could be developed by different feature teams. Ideally, features associated with one specific business division needed to be developed by people hired or paid directly from the respective division. Security was considered as important as scalability.

Some functionality had already been developed. For instance, one team created a solution for digitalizing service tickets – a feature that required a phone or fax machine beforehand. This team chose React with a Node.js service as their stack.

The teams responsible for the existing functionality were reassigned to form a new team for the portal project.

Team setup

A core team was tasked with forming a technical basis and establishing processes that needed to be followed by other teams. Three internal satellite teams were created with a dedicated business focus. More teams from other internal divisions, as well as hired externals, were accommodated once available.

The internal developers were all masters of **TypeScript** or were at least adept at writing JavaScript code. They all knew React – intermediately or better. Some external developers, however, were more familiar with Angular or other frameworks.

An overarching architecture guild and a dedicated testing team were available to bring alignment and ensure a high level of production quality. Importantly, there was a single solution architect who was allowed to make all the technical key decisions. This ensured consistency and fast reactions without much discussion.

The solution was brought to concept by the solution architect. Individual core pieces were distributed among the different teams to ensure knowledge sharing.

Let's see what the solution looked like.

Solution

The solution was based on three principles:

- Using a pattern library to establish design and behavior experience consistency with maximum reuse

- Developing reusable services that are combined dynamically in a **GraphQL** gateway used as a **BFF**

- Giving teams full autonomy to take ownership and responsibility, from testing up to production

For the sake of speed, some existing applications were not converted into micro frontends right away. Instead, links to these applications were placed in the portal. The applications only got a small facelift to make the experience consistent and – at least to the customer – look as if they were one solution.

In terms of the architecture for the micro frontend part, the siteless UI pattern was chosen. The app shell is responsible for providing translations, access to the gateway, routing, and dispatching events from the backend. All the micro frontend assets, as well as crucial app shell assets, are stored on a **CDN** to improve the throughput, allowing you to reach more end users and transfer the data faster.

The following diagram shows the essential architecture:

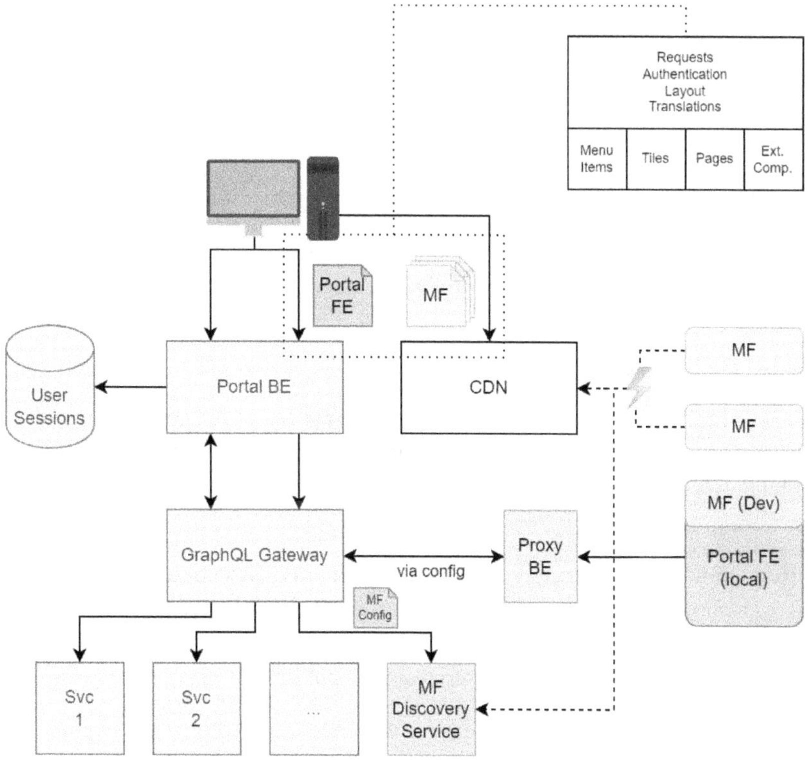

Figure 20.1 – Architecture of the micro frontend solution

One thing to note about the architecture, as laid out in *Figure 20.1*, is that each API request has to be done via the portal's backend. The reason for this architectural choice is that this component converts requests from cookie-based authorization into token-based authorization. The services running in the backend only understand the latter. When developing a micro frontend, this translation is done from a local backend proxy, which is started together with a special build of the portal's frontend. This includes the micro frontend, to enable things such as hot module reloading and standard dev tools.

Since most of the micro frontends were written using the React library, React and some related common libraries, such as React Router, were declared as shared dependencies in the portal. Also, the components from the pattern library were implemented using React.

The pattern library project was initiated beforehand, but it really got some drive during the portal project. It also made some design refinements and redesigns possible without much hassle. Since the pattern library was shared, most micro frontends remained the same, while the refinements were deployed with a new version of the app shell.

Some micro frontends were written in other frameworks. While most of them used the CSS classes from the pattern library, one micro frontend had a custom implementation for the styling. Consequently, the maintenance effort for this micro frontend was definitely greater when style adjustments were released.

The designs for the pattern library were created by a dedicated design team in parallel with the development team. While some screens were designed specifically to showcase the patterns and guide the way, most screens were left to the developers to follow the requirements provided by the product owner of each team.

Each team had one product owner, steered by an overarching chief product owner. For the UX and testing, shared resources and teams were set up. The architecture was also centralized.

With the project being first released after 4 months, the impact was expected to be meaningful.

Impact

The portal solution fulfilled its promise and enabled customers to access established and new digital services easily. It also provided a kind of lighthouse for other teams at the company on how to approach digital projects and incorporate modern concepts, such as API-first or a strong emphasis on modularization.

Before the introduction of the presented micro frontend solution, a technological patchwork carpet was seen, where customers had to use a lot of different websites and URLs to get information. These URLs were dependent on the market, business unit, and functionality. With the new solution, a single portal had this role – one that could be accessed from the public website.

Consequently, the company's public web presence was incorporated into the digital strategy. The micro frontend solution allowed its components to be used in other web applications, such as the CMS responsible for rendering the public website.

The portal could only be accessed by customers in supported regions and business segments. So, every user had to be approved first, which was usually done automatically. If approved, a customer could log in and get to the personal dashboard page shown in the following screenshot:

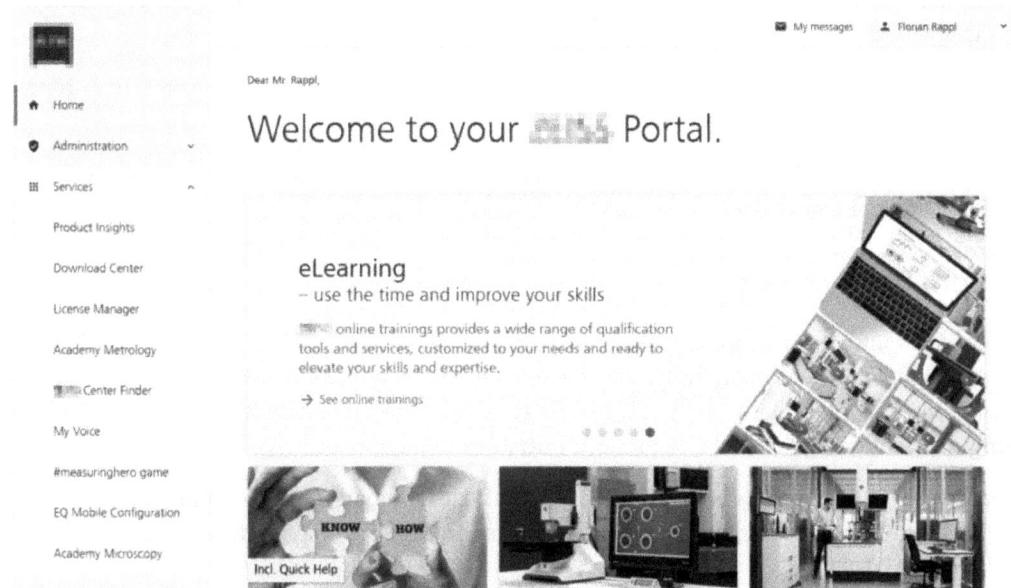

Figure 20.2 – The dashboard page of the customer portal

In the preceding screenshot, you can see the landing page of the portal for authenticated users. Each micro frontend could contribute tiles to the dashboard. All the menus were extensible and populated from the different micro frontends. For instance, the **My messages** button in the header is from the messages micro frontend. The user information itself is also a fragment of another micro frontend.

The portal allowed for progressive rollouts to be made to all desired markets. This enabled quick reactions to strategy adoptions and flexible, user-friendly feature integration. Another key factor of this was the ability to do autonomous development, where new business units could be onboarded quickly and without impacting existing work.

All in all, the solution shows that digital transformation with a micro frontend system works in a mostly in-house development-driven company – even without prior experience in or knowledge of distributed web applications.

Let's look at the second case study so that we can make a comparison.

An administration portal solution

In this second case study, you'll see how a German hidden champion introduced micro frontends for the administration of their new digital platform. The company specializes in developing and distributing high-quality tools and protective clothing, mostly for construction sites.

The company is organized as a single entity with over 70 subsidiaries all over the world. In Europe alone, the group has more than 440 sites.

Overall, it has around 4,000 employees worldwide. Its 2022 revenue was nearly one and a half billion euros. It operates in more than 50 countries.

Problem description

The company wanted to create an administration portal that allows full stack teams to provide a decent level of configurability to the administrators of a new digital service platform. A key idea was to keep the responsibility for the administration UI within the team of the respective API service, essentially not only distributing services but also the administration UI fragments.

The benefit of full stack teams is that the backend and the frontend move in alignment. As such, changing a feature on the backend puts the same team on the spot to align their change with the administration UI.

Team setup

An external core team was tasked with evaluating and presenting a potential technical basis. The same team then created the first version, which was refined until the first micro frontends could enter.

While the product owners were all internal, most developers and feature teams came from external companies. This alignment was always done via the company internally, so that each team could act as independently as possible.

A central testing team was available. Contract testing was necessary for backend APIs.

Let's see what the solution looked like.

Solution

The solution for this was quite similar to the previous case study, but different in the following areas:

- An existing pattern library had to be used – there were no dedicated designers for the project
- There was no GraphQL gateway, just some microservices that needed to be directly targeted
- Teams were not aligned anymore – the system was more of an open system, albeit only internally

The project started with a proof of concept to illustrate what the portal could look like and how development for the portal would be done. The scope for the proof of concept was to include identity management, which was one of the most sought-after functionalities of the admin portal. Also, this one contained sub-functionality such as adding or modifying identities, which requires more permissions than just reading identities. Therefore, this presented the opportunity to illustrate an extension system that allows different micro frontends to be loaded in accordance with the user's actual permissions.

To gain momentum, an established framework in the form of Piral was used. This introduced the portal's frontend as an emulator, which could be taken by micro frontend developers to debug and view their micro frontends before they were published. The app shell itself was maintained by the same team implementing the proof of concept.

Another difference to the previous case study was the CI/CD process. To accommodate the open system status, every team could set up their own CI/CD process. Ultimately, all these pipelines would lead to a common staging system, which was just an internal npm registry. Changes in the registry would be picked up by a common CI/CD system that was controlled by the owners of the platform. For the production and pre-production system, they would decide which micro frontends were published and when.

One of the reasons for picking this kind of release pipeline structure was to mimic the way that microservices are published on the same platform. This way, the different full stack teams did not have to learn anything new. The same pattern that they knew from deploying microservices packaged as Docker images was mirrored with micro frontends packaged as npm packages.

The similarities with the backend were also reflected in other choices. For instance, to ensure proper isolation while keeping the system stable, contract-based testing was introduced. The same system could then be leveraged for the frontend too, mainly to ensure that API calls remained stable. This was also done to help avoid introducing breaking changes in the portal's API, which could destroy some previously working micro frontends.

The architecture of the solution is outlined in the following diagram. Like in the previous case study, this solution involved a single architect reporting to a circle of architects. In this case, the circle used the proof of concept to decide how to continue with the project:

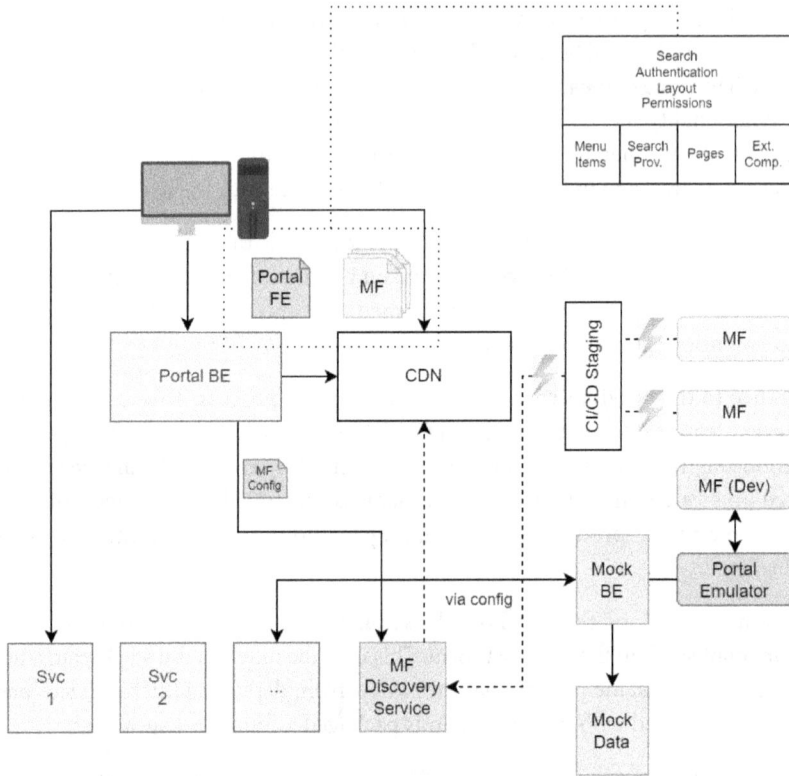

Figure 20.3 – Architecture of the micro frontend solution

In the architecture diagram of *Figure 20.3*, you can see that one of the app shell's responsibilities was to expose advanced search capabilities to the user. The search feature used a distributed approach where each micro frontend could contribute some additional search providers. When a global search started, all the available providers were called to contribute to the search results.

The search feature worked in three distinct modes:

- Global search request, which called all the available providers and aggregated the results
- Local search request, which only called the providers associated with the current page
- Local result filtering, which could be used on any page displaying tabular information

Consequently, the search feature of the admin portal was fluent and powerful. It gave the user access to information across all the available services quickly.

The flexibility and scalability of the admin portal also had quite an impact on the digital platform.

Impact

The admin portal helped the company achieve two things:

- Helping product owners, and admins in general, get directly involved in the platform
- Distributing the responsibility for providing administrative features to the different feature teams

It is fair to say that the solution fulfilled these expectations. Since the proof of concept was already received positively, the architecture circle decided to roll out the same approach for the customer portal. Later, parts of the public website were transformed, too.

For the standard – and other established – websites, a new micro frontend feed was introduced. The micro frontends in this feed did not use any shared dependencies and only solved certain common problems. One of these problems was presenting a common cookie consent form that unified the text, design, and behavior of the legally necessary cookie consent. With a single script, every website of the company could include the consent dialog, as well as other common functionality.

The following screenshot shows an early version of the admin portal:

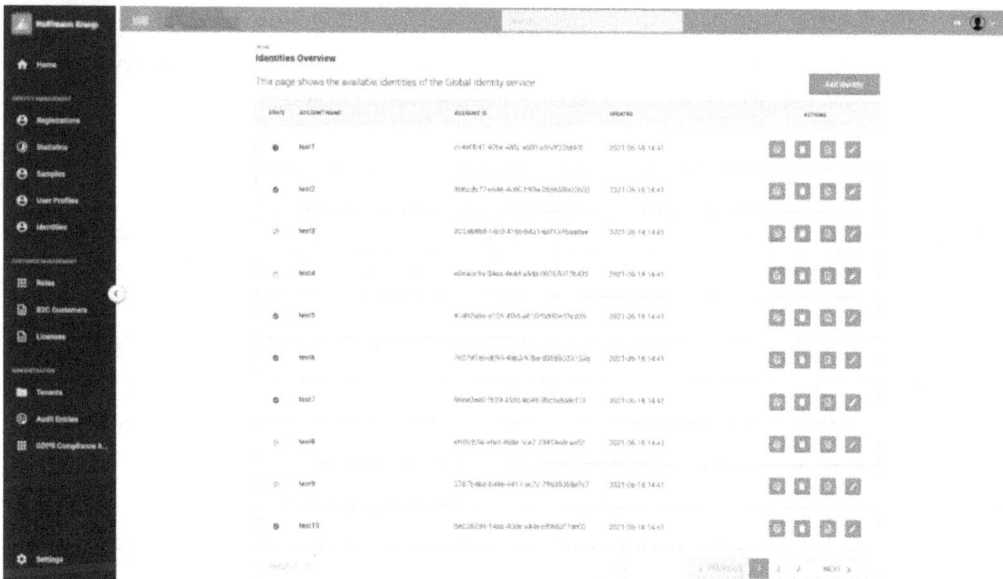

Figure 20.4 – The identities overview page of the admin portal

In the preceding screenshot, you can see that the page provides an overview of the managed identities. This page was provided by the `identity-management` micro frontend. The **Add identity** button, as well as most of the action buttons, are extensions that come from other micro frontends, such as the `create-identity` or `edit-identity` micro frontends. These micro frontends are only provisioned for users with the necessary backend privileges.

Using Piral as a basis was useful for leveraging the existing ecosystem and its documentation. This allowed the developers to focus on the domain-specific details in the documentation and the actual implementation. It also made onboarding new teams very fast.

As promised, the different teams took ownership of their provided functionality. If they changed something in the backend, they had the power and responsibility of also reflecting that change in the frontend. For the portal users, these changes had no impact as the whole solution just seemed to work coherently at all times.

Finally, it's time to look at the third case study, which deals with creating a very flexible web tool.

A healthcare management solution

OpenMRS is a collaborative open source project that helped create **electronic medical record (EMR)** software. It was first released in May 2014 and was originally written in Java.

OpenMRS, Inc. is a registered non-profit company that is the owner of all project-related intellectual property and the maintainer of the software's public license. The entity also represents the project in legal and financial matters.

Overall, the project has about 200 permanent contributors divided into a dozen teams. The development is mostly done on GitHub, with project management being centralized on Jira.

Problem description

The project's ecosystem was huge. While the backend provided a mechanism that could be extended in the form of Java modules, the frontend was often considered a necessary evil and was either reimplemented or replaced by a suitable alternative.

One problem here was that many distribution creators wanted to have a SPA experience. While some chose Angular as their framework, others went for React. There are even other variants out there. A micro frontend solution could help close this bridge by providing a modern SPA experience, along with a similar extensibility story on the backend.

Another aspect that needed to be considered was a rich set of configurability. The system needed to be extensible enough to be enriched with further capabilities, but sufficiently configurable to adjust everything from the look to the provided functionality.

Team setup

While each distribution had its own team, the core team was called the "micro frontend squad." In this squad, all fundamental technical decisions were discussed in the form of a **Request for Comments (RFCs)**. The squad worked in a geo-distributed fashion. Many developers worked in Africa, some in the USA, and others in Europe and India.

Much of the directed work was organized via JIRA. While some contributors were paid by different health organizations, others contributed as part of a larger fund or open source grant. Another fraction contributed their spare time.

Every week, all contributors gathered in a virtual dev. In this meeting, the vision and direction of the micro frontend project were discussed. There were demos and presentations of recent or upcoming work.

The micro frontend squad had one direct architect and several lead developers in different time zones – and from different organizations. While the team was centralized, the product owners were distributed. Other relevant resources, such as QA testers, were available within the OpenMRS project. These were shared via the different squads.

In the squad, almost every developer worked on a different item. Often, these items were also in different micro frontends, which gave the developers the advantage that no one stepped on someone else's toes. Still, some training was necessary as most developers had little to medium knowledge of TypeScript or hadn't used React, which was the framework of choice for most of the micro frontends maintained by the squad.

Let's see what the solution looked like.

Solution

The solution had to take care of the following points:

- No restrictions on frameworks or libraries
- Everything had to be configurable
- Easy to deploy and set up

The crucial point was decoupling the SPA from the API. Likewise, the micro frontends were set up in a way that they were given flexibility in terms of how they were distributed. For instance, while the SPA could be served directly from the OpenMRS backend via a dedicated module, it could also be served from a dedicated web server such as nginx. Similarly, the micro frontend assets could be embedded with the standard assets; however, they could also be hosted somewhere else.

In contrast to the previous two case studies, the solution did not dynamically provision the micro frontends. Instead, a static import map was used. The import map itself could, of course, be changed by hand. The micro frontends were rolled out in the ESM format supported by **SystemJS**.

The architecture of the solution is illustrated in the following diagram:

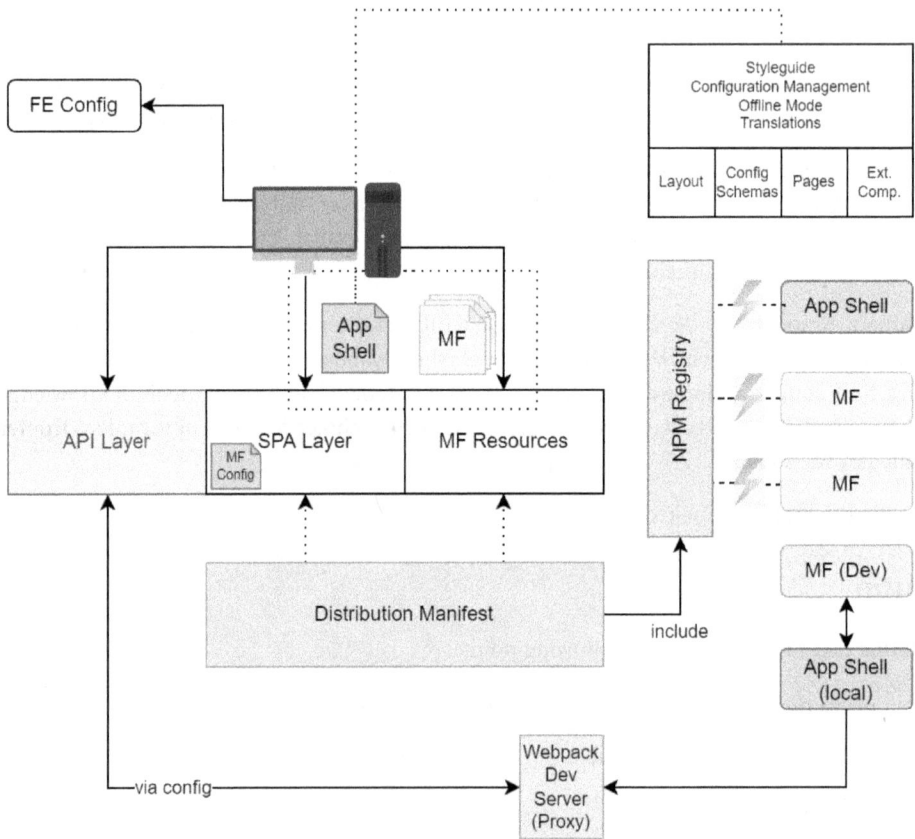

Figure 20.5 – Architecture of the micro frontend solution

OpenMRS is usually released in the form of different distributions. There is, however, also a reference application, which could be seen as the *default* distribution. In either case, the makers of a distribution can assemble all the necessary parts for the frontend via a distribution manifest. This concept is shown in the preceding diagram. The manifest defines what version of the app shell, and which micro frontends, should be included in the distribution.

All the resources are taken from npm packages that have been either released on the public npm registry or are available locally. This enables everyone in the community to publish their micro frontends independently and encourages micro frontend sharing.

To make things configurable, a frontend configuration system was introduced. System admins can specify a custom URL leading to a JSON file that overrides the default configuration. Consistency is ensured by checking the provided custom configuration against a defined configuration schema. Each micro frontend that offers configuration options must also declare a configuration schema.

The micro frontends are orchestrated by an app shell that leverages a single SPA for framework-independent routing. The extension slot mechanism is built on top of **single-spa** parcels. This was a good compromise between using an established technology with its ecosystem while keeping flexibility and fulfilling the custom requirements.

Initially, the app shell consisted almost exclusively of a single SPA root config and an import map. After identifying all cross-cutting concerns, some functionality was moved into the core. Additionally, some PWA fundamentals, such as a service worker, were integrated into the app shell to allow offline usage. All shared dependencies such as **React** and **Rxjs** were bundled into the application right away – only the micro frontends were present in the import map.

As far as the design is concerned, an existing solution was chosen: **Carbon Design** from IBM. The advantage of this choice is that it can be used with different frameworks and is feature-complete. It is a full pattern library and gives guidance on multiple screens. It also has a professional design that aligns with the design's vision and the ideas of the project.

But even with Carbon Design as a basis for the components, the designer took care of creating screens and crafting user flows. These flows were then tested together with some existing users to validate their usefulness and check what was still missing.

What was the impact of the freshly designed, fully dynamic, and modular application? We'll look at this in the next section.

Impact

Usually, novel solutions that try to replace established ones have a rough start. While some people eagerly wait for new solutions, most don't want to change their usage patterns. Also, the ones that are waiting are usually the ones with unrealistically high expectations.

In the case of the new OpenMRS frontend application, the community, overall, received it well. Since the classic UI hadn't been disabled and was still part of the default distribution, everyone could choose which application they wanted to use. This was also crucial as some less user-facing parts of the classic UI had not been ported over for the initial release.

The solution's biggest achievements have been its offline capabilities and framework independence. Its offline capabilities allow users to sync and take care of patients while not connected to the server. The framework's independence allows users to include pieces that have been developed by other organizations using their own stack. For instance, the standard form engine was implemented by a team that used Angular for rendering.

The following screenshot shows one of the most important screens of the application:

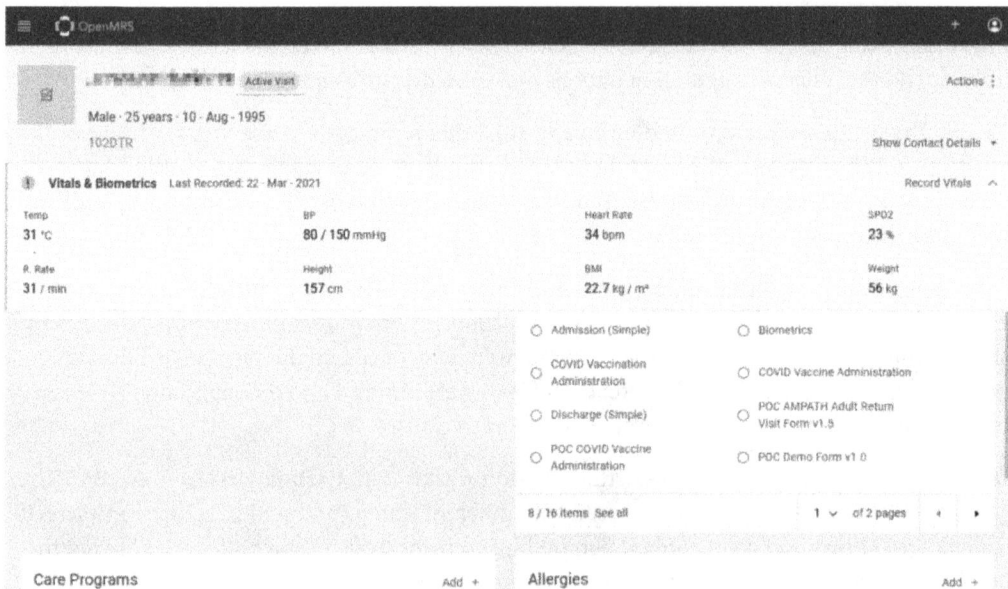

Figure 20.6 – The patient summary page with Vitals & Biometrics open

In the preceding screenshot, the patient summary page is displayed. This is the central dashboard for getting an overview of a patient. The summary page is just one of the many configurable and customizable dashboards. All dashboards are managed from a micro frontend called `patient-charts`. Different micro frontends, such as `patient-banner`, contribute to either all dashboards, such as by providing some UI fragment for the relevant patient information, or to special dashboards, such as the **Immunization** dashboard.

Since the flexibility of this approach comes with the downside of a quite complex and extensive configuration, a micro frontend for administrators and implementers was created: the implementor tools. This micro frontend allowed the current application to be inspected, modifications to be made, and the resulting configuration to be stored for reuse.

All in all, this solution unified the available solutions – instead of having each distribution come up with its own frontend, a single application now exists.

Let's look at a fourth case study to get more experience.

An e-commerce solution

For the fourth case study, we'll look at a company that created a **Software-as-a-Service (SaaS)** platform specializing in **workforce engagement management (WEM)**. The company is located in the US with about 300 employees. The revenue in 2022 was over 20 million USD. Their core products are quality assurance and workforce management. These products are complemented by features such as coaching, learning, and motivation – all tailored to enhance agent engagement across the employment lifecycle of contact center companies.

Problem description

The primary goal was to streamline user workflows and deliver value swiftly and securely. Achieving this required seamless integration of multiple products and a transition to an architecture enabling rapid iteration. While the backend and data infrastructure already leveraged serverless, microservices, and event-driven principles, the frontend posed a challenge. On the frontend, each team operated independently to meet evolving product requirements.

Team setup

The team structure did not include specialized teams for specific areas. Instead, within each team, the project had full stack developers who exhibited a certain level of specialization and proficiency in specific stacks, such as frontend development.

In total, the project had around 80 developers distributed across the world with hot spots in Colombia and Australia, collaborating within their teams to deliver high-quality solutions.

Architecture decisions were collaboratively made within and between the teams, respecting input from developers who had expertise in various areas. Initially, responsibility for micro frontends was assigned to a tribe team, consisting of senior developers from different teams. However, the resulting solutions demanded designating an owner, leading to the current architecture.

While no external architects were consulted, the teams relied on collective knowledge, references, books, meetups, talks, and online resources, as well as internal experience to back up their decisions, ensuring alignment with the project's goals.

Solution

The frontend stack exhibits a blend of diverse technologies and methodologies, reflecting the company's commitment to flexibility and innovation. By embracing micro frontends to modularize the frontend architecture, independent development and the deployment of individual features have been enabled for everyone.

For the actual implementation, Vue.js was chosen as the primary framework. The solution makes it possible to accommodate multiple versions, such as Vue 2, Vue 2.7, and Vue 3, suiting varying project requirements.

Orchestrating the micro frontends presented an intriguing challenge, which was addressed through a combination of edge-based routing using AWS services such as **CloudFront** and **lambda@edge**, as well as client-side routing facilitated by the `single-spa` library. This dual approach optimizes performance and maintains a seamless user experience across diverse environments. Choosing AWS services made sense as the backend was already using AWS as the cloud provider.

The architecture of the solution is illustrated in the following diagram:

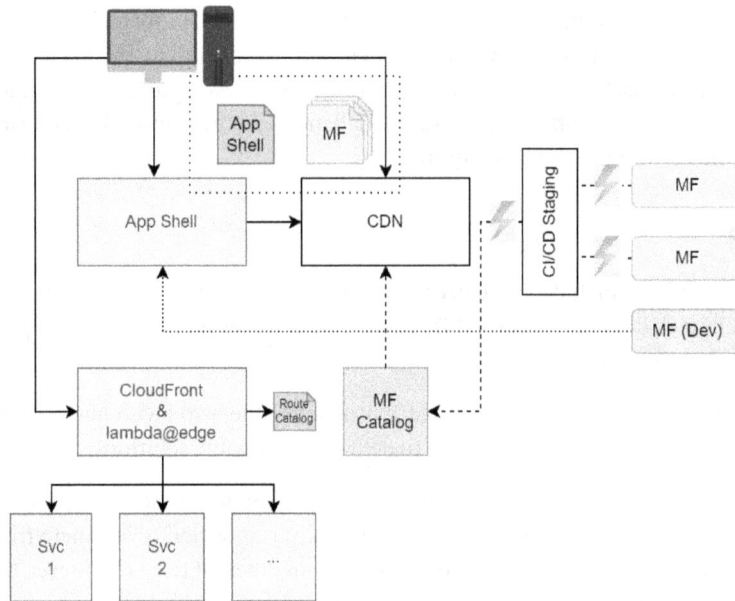

Figure 20.7 – Architecture of the micro frontend solution

In *Figure 20.7*, we see an architecture diagram of the project. Importantly, to develop micro frontends locally, we need to change the import map of the application to also include the locally available micro frontend. Another crucial aspect is that the frontend is now very similar to the backend – instead of a route catalog, we find a micro frontend catalog. Likewise, instead of CloudFront distributing the request, we use the `single-spa` package in the app shell. This symmetry makes it quite easy to work with the solution.

The deployment and distribution strategies are continuously evolving – for example, with plans to automate the creation of a JSON catalog for routing orchestration. This will enable smoother deployment processes and facilitate the implementation of advanced deployment strategies such as canary or blue-green deployments.

In terms of shared components and resources, the company leveraged web components developed using Lit and distributed via npm packages found in a private npm registry. This ensures consistency and reusability across the diverse frontend ecosystem. Overall, the frontend stack enabled the project to deliver scalable solutions while adapting to the dynamic requirements.

The journey from the initial draft to the go-live phase of the micro frontend's implementation was characterized by continuous iteration and evolution. While the process spanned approximately three years, it's important to note that this timeframe included multiple iterations and refinements. The overall goal was to strive for an optimized architecture that meets the evolving needs.

Impact

The implementation of micro frontends had a profound impact on the product's development process and user experience. By modularizing the frontend architecture, micro frontends enabled the developers to deliver faster updates and enhancements to the application, resulting in improved responsiveness and functionality for their users.

The primary goal to streamline the development and deployment processes was fulfilled. The company saw significant improvements in the CI/CD pipeline, with a unified approach to deploying frontend applications. This led to a faster time-to-market for new features and simplified deployment per view, enhancing the agility and responsiveness to user needs.

Furthermore, the micro frontend approach facilitated easier migration or adoption of new framework versions, allowing the developers to stay up to date with the latest technologies without disrupting the existing workflows. Additionally, the testing process underwent significant optimization. Thanks to import maps, testing is now easier, resulting in faster bug resolution and improved overall efficiency.

In terms of maintenance and continued development, the company remains committed to optimizing the micro frontend architecture by adding metrics, logs, and automation to drive ongoing improvements. Regular updates and enhancements are planned to ensure that the application continues to meet the evolving needs of the project's users and remains at the forefront of innovation in the industry.

Let's look at a fifth and final case study. To mix things up, we'll look at a solution that used micro frontends – but not to make a website. Instead, the following case study covers an application for mobile banking.

An application for mobile banking

The final case study is about an application for mobile banking in the context of JPMorgan Chase & Co. JPMorgan Chase & Co. is a US-based multinational finance company. The organization has around 300,000 employees and generated a little bit less than 160 billion USD of revenue in 2023. Besides banking, the company is also active in other financial areas such as risk management and private equity.

Problem description

The project had to be created without using any resources already available from the US branch of Chase. The basic idea was to have a mobile-only banking solution that could be introduced as Chase UK without demanding any physical infrastructure to be set up. No ATMs or branches would need to be established anywhere.

The app had to fulfill the security requirements provided by the Bank of England through the **Prudential Regulation Authority (PRA)**. As target platforms, only iOS and Android in more recent versions were determined.

Team setup

The distributed teams consisted of seasoned React developers, which required some getting used to React Native.

The core developers, as well as the product owners and decision makers, were all employed directly by Chase. Additional developers were included from external sources – all trusted companies that had been long-time partners in terms of providing development resources to Chase.

Solution

As a greenfield project, the team decided to go for React Native with TypeScript as the language of choice. The decision was made due to React Native's *write once, use everywhere* promise. Out-of-the-box React Native brings a monolithic architecture that makes it difficult to get micro frontends fully working. To circumvent this, the teams followed a feature-driven design methodology, grouping everything into features – which are then loosely coupled.

Loosely coupling these features was necessary to prevent all kinds of trouble. First of all, it made the solution much more flexible. Second, it also guaranteed an easier-to-understand code base. With strong coupling, a lot more shared dependencies would have been required. A first evaluation yielded that over 80 shared packages with a lot of inter-dependencies might be needed. This also would have introduced the danger of producing cyclic dependencies between these packages – eventually giving rise to hard-to-debug errors and high build times.

As far as the feature-driven design approach of the product teams is concerned, the main objective was to develop an overall model of the world before building a list of features, which were then planned, designed, and built by the respective teams. Shared npm packages were only introduced if their exported functionality was used across the board – and without relying on additional dependencies. This brought the number of shared packages down to only a handful.

Bringing down the number of shared packages was also crucial to optimize the CI/CD pipelines. The pipelines were built using Turborepo as the build system. This way, all artifacts could be stored and reused again – leading to cache hits and misses during a standard CI/CD pipeline run. In case of a cache hit, no testing or deployment step had to be run, while cache misses required these steps.

Previously introduced but outdated cache entries were removed from the system every three days with a cron job. This kept the data small and to the point without sacrificing much of the build performance.

The architecture of the solution is shown in the following diagram.

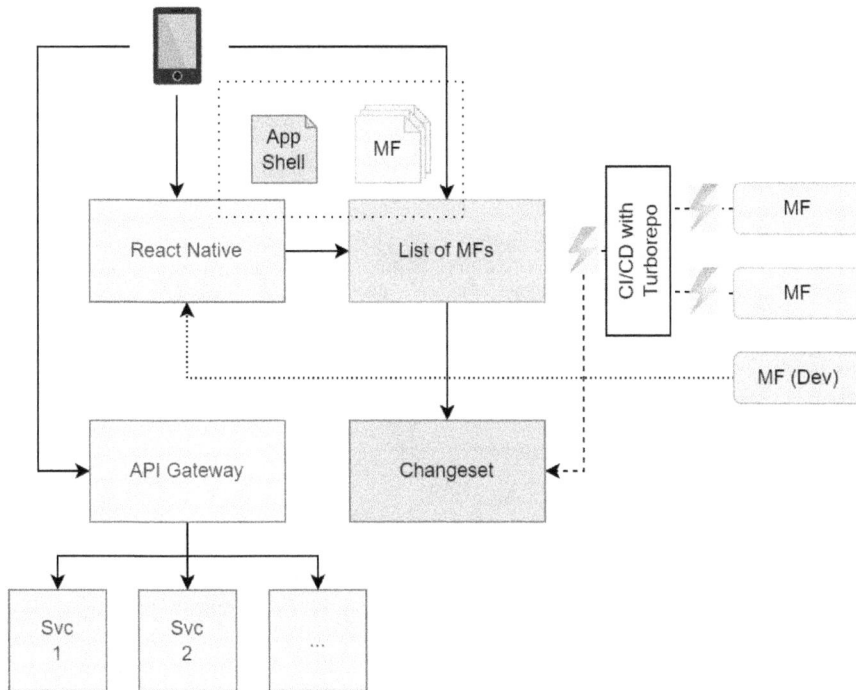

Figure 20.8 – Architecture of the micro frontend solution

In *Figure 20.8*, we see the architecture of the derived micro frontend solution. Thanks to the Metro bundler for React Native, the app is built with all micro frontends included in the changeset for the current release – all coming from a monorepo managed by **PNPM**. Alternative tools such as Repack were omitted due to their incompatibility with some of the tooling that was used.

Besides legal and technical implications, the main reason for bundling the micro frontends at compile time was the restrictiveness of the app stores. Both app stores, Apple's App Store and Google Play, explicitly forbid side-loading of additional code. This pretty much rules out runtime-composed micro frontends.

The version-specific changeset made it possible to only use the micro frontends in the version that had been enabled for them. This way, product owners could elevate the features in the version they'd like to see in a future release.

Impact

Today, the app is used by more than 2 million people – making it one of the fastest-growing banks in Europe. In the first two years since its release, the app went from zero to 1.6 million users, which is quite extraordinary for a non-challenger mobile-only bank in the UK.

The following screenshot illustrates how the app looks for end users:

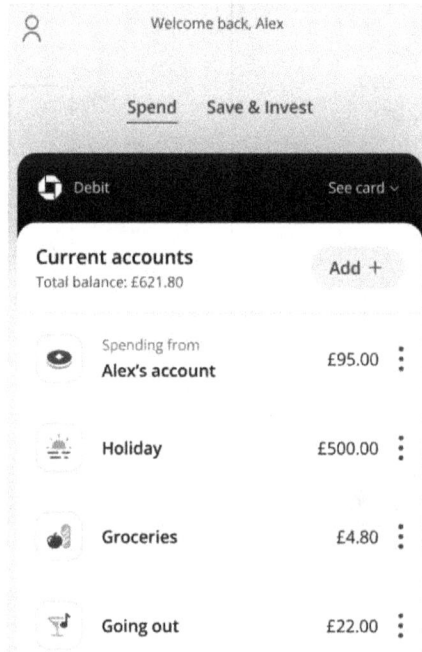

Figure 20.9 – The accounts feature of the app with an overview of spending

In *Figure 20.9*, we see the accounts feature of the app. Some of the other features include cards, rewards, insights, payments, and services. Each feature is developed by a separate team – with clear responsibility regarding the canvas usage and exposed UX.

The feature-driven design is very open to future extensions and has emerged as the right approach drawn from the experiences obtained in creating the app. After all, feature teams are quite autonomous, while still being part of the app as a whole. This fully conforms with regulations without sacrificing developer autonomy or flexibility.

The development of the app benefited mostly from the micro frontend architecture, with added flexibility due to autonomous teams, shortened build times, and improved CI/CD pipelines, as well as isolated defects and independent testing with subsequent releases. This made it easier to fail fast, which introduced some cost savings.

It's time to wrap up these case studies and look back on what you learned in the final chapter.

Summary

In this chapter, you learned how others approached the problem of introducing a micro frontend solution. You saw that the business factor always plays an important role in how technical decisions are made and what a micro frontend solution should look like. In contrast, teams being empowered by increased modularization is usually one of the key outcomes of a development experience.

Using micro frontends gives businesses a faster time to market, without having to wait for long release cycles and unrelated features to be implemented. If the domain is decomposed properly, then every module not only makes sense but can truly live on its own. Using patterns such as extension components, a UI can be composed dynamically and always adjust perfectly to the available functionality.

Micro frontends are as much a technological challenge as an organizational one. Communication with the stakeholders, as well as between the individual teams and different roles within a team, has to be changed too. You read about the need for change management, a solid governance model, dependency management, and design philosophy. Not all these aspects are needed right away, but they will come up and require some solution to be covered effectively. Following a pragmatic approach is certainly sufficient in most cases. Don't waste precious time and effort trying to come up with the perfect solution.

Epilogue

With this, your journey into micro frontends ends. Now, it's time to use your knowledge and build outstanding web applications using the techniques, tips, tricks, and practices shown in this book. I hope you gained some crucial insights and enjoyed reading about how to decouple components. Creating independent modules will only work if the necessary groundwork is done – independent if the foundational architecture fully qualifies as micro frontends or some intermediate pattern such as a modulith. Coming up with a good domain decomposition is always difficult, so don't be alarmed when the first shot needs additional refinement to work decently.

Also, don't forget to evaluate whether a micro frontend solution is really needed. A simple monolith cannot be beaten in terms of effectiveness and performance. Choosing micro frontends should be less about the technical challenge and more about the actual business need. Be open in your communication and try to listen to everyone before making a decision in favor of or against micro frontends in general, or a particular architecture style.

All the best!

Index

A

A/B testing 16, 17
administration portal solution 300
  impact 303, 304
  problem description 300
  solution, to problem 300-302
  team setup 300
Angular 217
anti-forgery token 268
application, for mobile banking 311
  impact 314, 315
  problem description 312
  solution, to problem 312-314
  team setup 312
application programming
    interface (API) 6, 37, 53
app shell 128
architectural boundaries 54
  DOM, accessing 57, 58
  right level of freedom, selecting 56
  shared capabilities 55
  universality of micro frontends 58
Astro 110
asynchronous JavaScript and XML (AJAX) 5
Asynchronous Module Definition
    (AMD) 208

authenticity
  verifying 229, 230
Azure DevOps 31

B

backend-driven micro frontends
  versus frontend-driven micro
    frontends 70-73
Backend For Frontend (BFF) 25, 88, 297
  stitching 124
best developer experience
  achieving 288
  customization, via browser extensions 292
  developer portal, implementing 292-294
  error codes, integrating 289, 290
  offline-first development environment,
    providing 290, 291
Bit
  URL 66
Blazor 217
blue-green deployments 240
blue micro frontend 93
bootstrap function 156
Bootstrap library 218
bounded context 45, 46
Browsersync 290

bundlephobia
    URL 22

C

Carbon Design 307
Cascading Style Sheets (CSS) 5
central dependency sharing 199
central deployments 32
    mono repository (monorepo), using 32-35
    multiple repositories, joining 35, 36
central linking directory 81
central user management 23
change management 259, 260
CircleCI 31
    used, for multiple repositories 37
C-level stakeholders
    communicating 246
    executive summaries, writing 248
    expectations, managing 246-248
client-side composition 128
    architecture 128-130
    potential enhancements 135
    pros and cons 135
    sample implementation 130-134
client-side rendering (CSR) 5, 70
closed mode 137
CloudFront 310
Code in Action (CiA) 246
cold reload 139
Common Gateway Interface (CGI) 4
communication patterns
    components, extending 162
    data sharing 161
    events, exchanging 160, 161
    exploring 159
composition layout
    creating 102
    ESI, using 104

JS template strings, using 104, 106
    responsibilities 103
    SSI, using 103
Content Delivery Network (CDN) 129, 297
Content Security Policy (CSP) 57, 224
context map 46
continuous integration (CI) 31
**Cross-Origin Resource Sharing
    (CORS)** 12, 226
Cross-Site Request Forgery (CSRF) 268
Cross-Site Scripting (XSS) 224, 265
CSS
    implementation techniques,
        to scope 213-217
    modern features, using for isolation 219-222
CSS-in-JS 215
cURL 177

D

DDD principles 44
    bounded context 45, 46
    context map 46
    modules 44
    strategic domain design, versus
        tactical design 46, 47
decent developer experience
    code analysis, assisting with 287
    code documentation, centralizing 285, 286
    establishing 285
    testability, improving 288
    videos, using 286
dedicated rendering server
    using 108, 109
Deno Fresh 110
Dependabot 268
dependencies
    sharing 258, 259
    sharing, between micro frontends 198-200

discovery service
  advanced capabilities, adding 237-240
  implementing 234-237
distributed deployments 37
  dedicated pipelines, using 39
  monorepo, using 37-39
distributed sharing 199
Docker Compose 117
Document Object Model (DOM) 11, 54
  accessing 57
Docusaurus
  URL 285
Docz 277
domain-driven design (DDD) 15, 43
  principles 44
Domain Specific Language (DSL) 278
dynamic micro frontends
  versus static micro frontends 65-68

E

ECMAScript 5 (ES5) 58
e-commerce solution
  impact 311
  problem description 309
  solution, to problem 309-311
  team setup 309
edge-side composition 112
  advantages 118-120
  architecture 112, 113
  disadvantages 118-121
  potential enhancements 117, 118
  sample implementation 113-117
Edge-Side Includes (ESI) 71, 91, 122-124
  using 104
electronic medical records (EMR) 304
emerging web standards 10
  frame communication 11, 12
  isolation, via web components 11

  proxies 13
  web workers 13
Emotion library 215
ESLint 267, 287
ES Modules (ESM) 154, 207
Extensible Markup Language (XML) 6
extension slot 162

F

faster TTM 14
  A/B testing 16
  isolated features 16, 17
  multiple teams 15
  onboarding time, decreasing 14, 15
feed of micro frontends 169
feed server 170
First Contentful Paint (FCP) 56
flat stitching approach 124
fragments 83
  using, with iframes 83
frame communication 11
framework-agnostic components
  implementing 188, 189
frontend, and backend
  separating 6, 7
frontend-driven micro frontends
  versus backend-driven micro
    frontends 70-73
functional split 49, 50, 51
Function-as-a-Service (FaaS) 106, 179

G

**General Data Protection
    Regulation (GDPR) 266**
generic layout 103
getModuleContent function 171

getStream function 192
governance model
  establishing 261-263
GraphQL 23
GraphQL gateway 297
green micro frontend 94

H

hash algorithm 230
hash value 230
healthcare management solution 304
  impact 307, 308
  problem description 304
  solution, to problem 305-307
  team setup 304, 305
hidden monoliths
  avoiding 232, 233
horizontal split micro frontends
  versus vertical split micro frontends 69, 70
hot reload 139
http-proxy-middleware 290
HTTPS 12
HTTP Strict Transport Security (HSTS) 226
hybrid solutions 40
  release on change, triggering 40, 41
  scheduled releases 40
hyperlinks 81
  using, for navigation 81
HyperText Markup Language 5
    (HTML5) specification 4
HyperText Markup Language (HTML) 58
HyperText Transfer Protocol (HTTP) 37, 57

I

iframes, challenges
  accessibility 83
  layout 83, 84

security 83
import maps 155
individual user management 24
inline frame (iframe) 4
  fragments, using with 83
Internet Information Services (IIS) 5
islands architecture 190
islands composition 189
  implementing, with Qwik 190-194
  with siteless UIs 191
islands of interactivity 109
isolation
  via web components 11

J

JavaScript Object Notation (JSON) 6
JavaScript Web Tokens (JWTs) 25
jsdoc comments 287
JSHint 267
JSON Web Token (JWT) 224
JS template strings
  using 104-106

K

knowledge sharing 26-28
kras 290

L

lambda@edge 310
Lerna 33
  URL 33
LGTM 267
local linking directory 82

M

man-in-the-middle (MITM) 57, 226
micro frontend 2 (MF2) 233
micro frontend projects 106
  composing 138
  discovery service, using 138, 139
  lifecycle, examining 108
  podlets 107
  setting up 106
  updating, at runtime 139, 141
micro frontends 10
  backend-driven, versus frontend-
    driven 70-73
  dependencies sharing, between 198-200
  horizontal, versus vertical split 69, 70
  static, versus dynamic 65-68
micro frontends, case studies
  administration portal solution 300
  application for mobile banking 311
  ecommerce solution 309
  healthcare management solution 304
  user-facing portal solution 295, 296
micro frontends landscape 64, 65
  3D phase space 65
micro frontend squad 304
microservices 10
  advantages 8
  disadvantages 9
minimum developer experience
  development, support in
    standard IDEs 282, 283
  providing 282
  scaffolding experience, improving 283-285
minimum viable product (MVP) 16
modern hydration
  optimizing 163, 164
ModHeader 292
module 44

Module Federation 68, 155, 200
  utilizing 200-205
mono repository (monorepo) 66
  using 32-39
Mosaic 9 71
  reviewing 101
mount function 156
multiple repositories
  joining 35, 36
  with CircleCI 37

N

Native Federation 205-207
nodejsscan 267
non-disclosure agreements (NDAs) 15
npm 33
npm package format 181
npm packages 129
npm registry 66
Nx 33

O

OpenAPI specification 290
open mode 137
OpenMRS 278, 304
OpenMRS, Inc. 304
open styling
  consequences 212, 213
Open Web Application Security
  Project (OWASP) 265

P

performance 20
  bundle size 21, 22
  request optimizations 22
  resource caching 20, 21

PHP: Hypertext Preprocessor (PHP) 5
Picard 205
pilets 184
pilet's life cycle 185-187
    functional life cycle 185
    software development life cycle 185
Piral 70, 284
    used, for building siteless UI
        runtime 182-184
    used, for deploying siteless UI
        runtime 184, 185
Piral Cloud Feed service
    URL 234
Piral Inspector 292
PNPM 268, 313
Podium 69, 101, 102
    layout service 101
    podlet 101
    URL 102
podlets 107
policy
    testing 225
polyfill 138
private npm registry 66
product owners (POs) 16
    handling 249, 250
progressive rendering 190
progressive rendering techniques 164, 165
proof of concept (POC) 4
prop drilling 177
proxies 13
Prudential Regulation Authority (PRA) 312
public npm registry
    reference link 66

Q

Qwik 110
    used, for implementing islands
        composition 190-194

R

React 307
red micro frontend 90
registerApplication function 152
reliability 28
removeLoader parameter 193
representational state transfer (REST) 7
Request for Comments (RFCs) 304
Responsible - Accountable - Consulted
        - Informed (RACI) 249
resumability 190
root config 152
rspack 200
Rxjs 307

S

sandboxing micro frontends 263, 264
script access
    limiting 228, 229
search engine optimization (SEO) 83
Secure Hash Algorithms (SHA) 230
security 23
    central user management 23
    hardening, with web standards 224-227
    individual user management 24
    script execution 25, 26
security concerns 265-268
separation of concerns (SoC) 43, 48
    example decomposition 51-54
    functional split 49-51
    technical split 48, 49

serverless
versus siteless UIs 180
server-side composition
advantages 100
architecture 89
basics 88
disadvantages 100
gateway, implementing 96-99
known users 102
list of recommendations,
    implementing 94, 95
potential enhancements 100
products page, implementing 90-92
sample implementation 89, 90
store functionality, implementing 93, 94
Server-Side Includes (SSI) 5, 71
using 103
server-side rendering (SSR) 4, 20, 65, 70
service-oriented architectures (SOAs) 7, 64
shadow DOM 11, 135
closed mode 137
open mode 137
used, for isolating styles 137, 138
using 217-219
sharing hints 204
single-page applications
    (SPAs) 6, 20, 143, 290
single-responsibility principle (SRP) 8
single-spa-inspector 292, 307
single-spa library 73
single-spa package 152
siteless UI modules
framework-agnostic components,
    implementing 188, 189
pilet's life cycle 185-187
writing 185

siteless UI runtime
building, with Piral 182-184
creating 181, 182
deploying, with Piral 184, 185
siteless UIs 168
advantages 179
architecture 168, 169
developing, locally 180
disadvantages 179
implementation 170
modules, publishing 181
potential enhancements 178
versus serverless 179
siteless UIs, implementation
app shell 172-174
balance micro frontend 176, 177
feed server 170-172
settings micro frontend 175, 176
tax micro frontend 174, 175
social web 5, 6
Software-as-a-Service (SaaS) 59, 309
SonarCloud 267
SonarSource 267
SPA composition 144
advantages and disadvantages 150, 151
architecture 144, 145
implementation 145-150
potential enhancements 150
SPA micro frontends
cross-framework components,
    using 157-159
integrating 156, 157
lifecycle, declaring 156
SPA shell
building 151
dependencies, sharing 153-155
pages, activating 151-153
specific layout 103

SSI 121, 122

SSRStream component 192

standard IDEs 282

static micro frontends

versus dynamic micro frontends 65-68

steering committees 249, 250

Storybook 277

strategic domain design

versus tactical design 46-48

Styled Components 215

Styleguidist 277

style isolation 137

with shadow DOM 138

StyleX 215

Subresource Integrity (SRI) 57, 229

swarming 249

SystemJS 154, 208, 305

independence, achieving with 208, 209

T

tactical design

versus strategic domain design 46, 47

Tailwind CSS

reference link 264

teamchilla Blueprint 283

team organization 250

changing 254

team setup 251-254

technical split 48, 49

testability 288

The European Organization for
Nuclear Research 4

three-dimensional (3D) phase space 64

Time-To-Interactive (TTI) 56

TSHint 267

TypeScript 287, 296

typings 287

U

Uniform Resource Locators (URLs) 68

handling 4

Universal Router

reference link 134

unload function 156

unmount function 156

User Experience (UX) 257, 269

pattern libraries 29

wording 29

user-facing portal solution 295

impact 298, 299

problem description 296

solution, to problem 297, 298

team setup 296

user interface (UI) technologies 64

UX design

creating, without designers 278

elements, adding 270-273

sharing 276, 277

starting, with empty application
shell 274-276

V

Vanilla Extract 215

Varnish 104

URL 104

Varnish Configuration Language (VCL) 104

Verdaccio

URL 66

vertical split micro frontends

versus horizontal split micro
frontends 69, 70

Visual Studio Code (VS Code) 64

Vite bundler 200

Vue 217

Vuepress 277

W

W3C, ESI 1.0
reference link 122
web
programming 4
web applications
evolution 4
web approach
advantages 80
architecture 77
basics 76
disadvantages 80
hyperlinks, using for navigation 81
implementation 78
potential enhancements 80
sample implementation 77-79
web components 136, 137
custom elements 136
HTML templates 136
shadow DOM 136
styles, isolating with shadow DOM 137, 138
web standards
using, to harden security 224-227
web workers 13
**workforce engagement management
(WEM) 309**

Y

**YAML Ain't Markup Language
(YAML) files 31**
Yarn 33, 268
Yeoman 283

Z

Zed Attack Proxy (ZAP) 263
Zeplin 270
reference link 270

‹packt›

Subscribe to our online digital library for full access to over 7,000 books and videos, as well as industry leading tools to help you plan your personal development and advance your career. For more information, please visit our website.

Why subscribe?

- Spend less time learning and more time coding with practical eBooks and Videos from over 4,000 industry professionals

- Improve your learning with Skill Plans built especially for you

- Get a free eBook or video every month

- Fully searchable for easy access to vital information

- Copy and paste, print, and bookmark content

Did you know that Packt offers eBook versions of every book published, with PDF and ePub files available? You can upgrade to the eBook version at packtpub.com and as a print book customer, you are entitled to a discount on the eBook copy. Get in touch with us at customercare@packtpub.com for more details.

At www.packtpub.com, you can also read a collection of free technical articles, sign up for a range of free newsletters, and receive exclusive discounts and offers on Packt books and eBooks.

Other Books You May Enjoy

If you enjoyed this book, you may be interested in these other books by Packt:

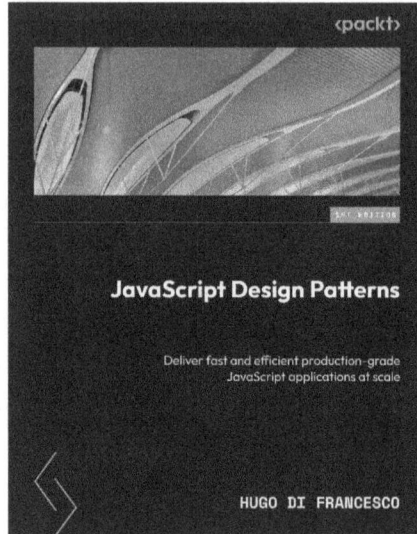

JavaScript Design Patterns

Hugo Di Francesco

ISBN: 978-1-80461-227-9

- Find out how patterns are classified into creational, structural, and behavioral
- Implement the right set of patterns for different business scenarios
- Explore diverse frontend architectures and different rendering approaches
- Identify and address common asynchronous programming performance pitfalls
- Leverage event-driven programming in the browser to deliver fast and secure applications
- Boost application performance using asset loading strategies and offloading JavaScript execution

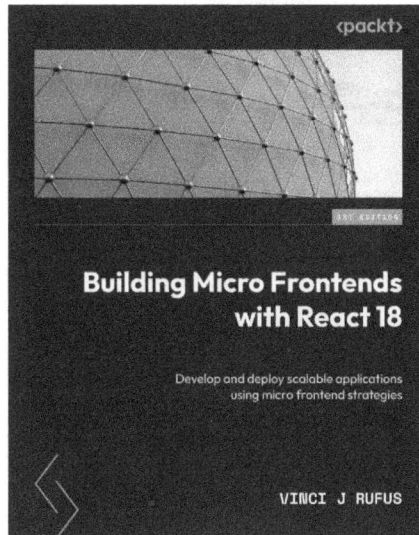

Building Micro Frontends with React 18

Vinci J Rufus

ISBN: 978-1-80461-096-1

- Discover two primary patterns for building micro frontends
- Explore how to set up monorepos for efficient team collaboration
- Deal with complexities such as routing and sharing state between different micro frontends
- Understand how module federation works and use it to build micro frontends
- Find out how to deploy micro frontends to cloud platforms
- Figure out how to build the right development experience for teams

Packt is searching for authors like you

If you're interested in becoming an author for Packt, please visit authors.packtpub.com and apply today. We have worked with thousands of developers and tech professionals, just like you, to help them share their insight with the global tech community. You can make a general application, apply for a specific hot topic that we are recruiting an author for, or submit your own idea.

Hi!

I am Florian Rappl, author of *The Art of Micro Frontends (2nd Edition)*. I really hope you enjoyed reading this book and found it useful for increasing your productivity and efficiency.

It would really help me (and other potential readers!) if you could leave a review on Amazon sharing your thoughts on this book.

Go to the link below or scan the QR code to leave your review:

`https://packt.link/r/1835460356`

Your review will help us to understand what's worked well in this book, and what could be improved upon for future editions, so it really is appreciated.

Best wishes,

Florian Rappl

Download a free PDF copy of this book

Thanks for purchasing this book!

Do you like to read on the go but are unable to carry your print books everywhere?

Is your eBook purchase not compatible with the device of your choice?

Don't worry, now with every Packt book you get a DRM-free PDF version of that book at no cost.

Read anywhere, any place, on any device. Search, copy, and paste code from your favorite technical books directly into your application.

The perks don't stop there, you can get exclusive access to discounts, newsletters, and great free content in your inbox daily

Follow these simple steps to get the benefits:

1. Scan the QR code or visit the link below

https://packt.link/free-ebook/978-1-83546-035-1

2. Submit your proof of purchase
3. That's it! We'll send your free PDF and other benefits to your email directly

www.ingramcontent.com/pod-product-compliance
Lightning Source LLC
Chambersburg PA
CBHW061800210326
41599CB00034B/6819